Essential Advanced Precalculus

A Self-Teaching Guide

Tim Hill

Essential Advanced Precalculus: A Self-Teaching Guide
by Tim Hill

Copyright © 2018 by Questing Vole Press. All rights reserved.

Editor: Kevin Debenjak
Proofreader: Diane Yee
Compositor: Kim Frees
Cover: Questing Vole Press

Contents

1 **Sets** 1
   Elements and Cardinality 1
   The Cartesian Product 5
   Subsets 8
   Power Sets 12
   Union, Intersection, and Difference 14
   Complement 16
   Venn Diagrams 17
   Indexed Sets 20
   The Well-Ordering Principle 25
   Russell's Paradox 27

2 **The Real Number System** 35
   Properties of Numbers 35
   Absolute Value 47
   Natural Numbers and Induction 49
   Summation 53
   Integers, Rationals, and Reals 55

3 **Functions** 65
   The Concept of a Function 65
   Notation and Domains 67
   Function Types and Composition 70
   Formal Definitions 74
   Ordered Pairs 77

4 Graphs **83**
  The Real Line and Intervals 83
  Coordinates, Slope, and Distance 86
  Linear Functions 89
  Parabolas and Power Functions 91
  Polynomial Functions 93
  Rational Functions 94
  Oscillating Functions 95
  Ellipses and Hyperbolas 99

5 Solutions **111**

Index **165**

1 Sets

Elements and Cardinality

A **set** is a collection of entities, called the **elements** of the set. We're concerned with sets whose elements are mathematical objects, such as numbers, functions, points, and so on.

A set is often expressed as a list of comma-separated elements enclosed by braces. The collection $\{1, 3, 5, 7\}$, for example, is a set that contains four elements: the numbers 1, 3, 5, and 7. A set can have infinitely many elements; the collection of all integers, for example, is

$$\{\ldots, -3, -2, -1, 0, 1, 2, 3, \ldots\}$$

Here the dots indicate a pattern of numbers that continues endlessly in both the positive and negative directions. A set that contains infinitely many elements is an **infinite** set; otherwise, it's a **finite** set.

Two sets are **equal** if they contain exactly the same elements. Hence $\{1, 3, 5, 7\} = \{3, 1, 7, 5\}$ because the elements are identical, regardless of order; but $\{1, 3, 5, 7\} \neq \{1, 3, 5, 6\}$. Also

$$\{\ldots, -3, -2, -1, 0, 1, 2, 3, \ldots\} = \{0, -1, 1, -2, 2, -3, 3, \ldots\}$$

Uppercase letters are often used to denote sets. We can define $A = \{1, 3, 5, 7\}$, for example, and then use A to refer to $\{1, 3, 5, 7\}$. To state that 3 is an element of the set A, we write $3 \in A$, which is read as "3 is an element of A", or "3 is in A", or simply "3 in A". We also have $1 \in A$, $5 \in A$, and $7 \in A$, but $4 \notin A$. This last expression is read "4 is not an element of A", or "4 not in A". To indicate that several elements are in a set, use expressions like $5, 1 \in A$ or $1, 3, 7 \in A$.

Some sets are so important and prevalent that special symbols are reserved for them. The set of **natural numbers** (that is, the positive whole numbers) is denoted by $\mathbb{N}$:

$$\mathbb{N} = \{1, 2, 3, 4, 5, \ldots\}$$

The set of **integers** is another fundamental set:

$$\mathbb{Z} = \{\ldots, -3, -2, -1, 0, 1, 2, 3, \ldots\}$$

The set of all **real numbers** (that is, rational and irrational numbers) is denoted by the symbol $\mathbb{R}$. Other special sets are listed later in this section.

Sets needn't have only numbers as elements:

- The set $B = \{T, F\}$ consists of two letters, perhaps representing the values "true" and "false".

- The set $C = \{a, e, i, o, u\}$ consists of the lowercase vowels in the English alphabet.

- The set $D = \{(0, 0), (1, 0), (0, 1), (1, 1)\}$ has as elements the four corner points of a square on the x–y coordinate plane. Thus $(0, 0) \in D$, $(1, 0) \in D$, and so on, but $(1, 2) \notin D$ (for example).

Sets can also contain other sets as elements. For example, the set $E = \{1, \{2, 3\}, \{2, 4\}\}$ has three elements: the number 1, the set $\{2, 3\}$, and the set $\{2, 4\}$. Thus $1 \in E$ and $\{2, 3\} \in E$ and $\{2, 4\} \in E$. But note that $2 \notin E$, $3 \notin E$, and $4 \notin E$.

The set $M = \left\{ \begin{bmatrix} 0 & 0 \\ 0 & 0 \end{bmatrix}, \begin{bmatrix} 1 & 0 \\ 0 & 1 \end{bmatrix}, \begin{bmatrix} 1 & 0 \\ 1 & 1 \end{bmatrix} \right\}$ contains three two-by-two matrices. We have $\begin{bmatrix} 0 & 0 \\ 0 & 0 \end{bmatrix} \in M$, but $\begin{bmatrix} 1 & 1 \\ 0 & 1 \end{bmatrix} \notin M$. Letters can serve as symbols denoting a set's elements: if $a = \begin{bmatrix} 0 & 0 \\ 0 & 0 \end{bmatrix}$, $b = \begin{bmatrix} 1 & 0 \\ 0 & 1 \end{bmatrix}$, and $c = \begin{bmatrix} 1 & 0 \\ 1 & 1 \end{bmatrix}$, then $M = \{a, b, c\}$.

If X is a finite set, then its **cardinality** or **size**, denoted by $|X|$, is the number of elements that it contains. Hence for the sets above, $|A| = 4$, $|B| = 2$, $|C| = 5$, $|D| = 4$, $|E| = 3$, and $|M| = 3$.

An important set is the collection that contains no elements at all, the **empty set** or **null set**, denoted by $\emptyset$. Although it may not strike you as a collection in the ordinary sense of the word, the empty set arises naturally in many contexts. We frequently consider the set K, consisting of all x satisfying some property P. Often we have no guarantee that P is satisfied by any number, so that K might be $\emptyset$—in fact, often one proves that P is always false by showing that $K = \emptyset$.

Observe that $|\emptyset| = 0$. The empty set is the only set whose cardinality is zero. Be careful in writing the empty set. Don't write $\{\emptyset\}$ when you mean $\emptyset$. These sets aren't equal because $\emptyset$ contains nothing whereas $\{\emptyset\}$ contains one member, namely the empty set. If this concept confuses you, think of a set as a box containing things, so, for example, $\{1, 3, 5, 7\}$ is a "box" containing four numbers. The empty set $\emptyset = \{\}$ is an empty box. By contrast, $\{\emptyset\}$ is a box with an empty box inside it. Clearly, they're different: an empty box isn't the same as a box with an empty box inside it. Thus $\{\emptyset\} \neq \emptyset$. Also note that $|\emptyset| = 0$ and $|\{\emptyset\}| = 1$ as additional evidence that $\{\emptyset\} \neq \emptyset$.

The set $F = \{\emptyset, \{\emptyset\}, \{\{\emptyset\}\}\}$ might look odd but it's actually a simple set. Think of it as a box containing three things: an empty box, a box containing an empty box, and a box containing a box containing an empty box. Thus $|F| = 3$. The set $G = \{\mathbb{N}, \mathbb{Z}\}$ is a box containing two boxes, the box of natural numbers and the box of integers. Thus $|G| = 2$.

Set-builder notation is used to describe sets that are too big or complex to list between braces. Consider the infinite set of even integers $E = \{\ldots, -6, -4, -2, 0, 2, 4, 6, \ldots\}$. In set-builder notation this set is written as

$$E = \{2n : n \in \mathbb{Z}\}$$

The first brace is read "the set of all elements of form", and the colon is read "such that". So the expression $E = \{2n : n \in \mathbb{Z}\}$ is read as "E equals the set of all elements of form $2n$, such that n is an element of $\mathbb{Z}$". In other words, E consists of all possible values of $2n$, where n takes on all values in $\mathbb{Z}$.

In general, a set X written with set-builder notation has the syntax

$$X = \{\text{expression} : \text{rule}\}$$

where the elements of X are understood to be all values of "expression" that are specified by "rule". The set E above, for example, is the set of all values of the expression $2n$ that satisfy the rule $n \in \mathbb{Z}$.

This notation lets us express the same set in different ways without ambiguity. For example, $E = \{2n : n \in \mathbb{Z}\} = \{n : n \text{ is an even integer}\} = \{n : n = 2k, k \in \mathbb{Z}\}$. Another common way of writing this set is

$$E = \{n \in \mathbb{Z} : n \text{ is even}\}$$

which is read "E is the set of all n in $\mathbb{Z}$ such that n is even." Some writers use a vertical bar instead of a colon; for example, $E = \{n \in \mathbb{Z} \mid n \text{ is even}\}$.

Here are some more examples of set-builder notation:

1. $\{n : n \text{ is a prime number}\} = \{2, 3, 5, 7, 11, \ldots\}$
2. $\{n \in \mathbb{N} : n \text{ is a prime number}\} = \{2, 3, 5, 7, 11, \ldots\}$
3. $\{n^2 : n \in \mathbb{Z}\} = \{0, 1, 4, 9, 16, 25, \ldots\}$
4. $\{x \in \mathbb{R} : x^2 - 2 = 0\} = \{\sqrt{2}, -\sqrt{2}\}$
5. $\{x \in \mathbb{Z} : x^2 - 2 = 0\} = \varnothing$
6. $\{x \in \mathbb{Z} : |x| < 4\} = \{-3, -2, -1, 0, 1, 2, 3\}$
7. $\{2x : x \in \mathbb{Z}, |x| < 4\} = \{-6, -4, -2, 0, 2, 4, 6\}$
8. $\{x \in \mathbb{Z} : |2x| < 4\} = \{-1, 0, 1\}$

Examples 6–8 above highlight a conflict of notation: the expression $|X|$ means *absolute value* if X is a number or *cardinality* if X is a set. This distinction is always clear from context. Consider $\{x \in \mathbb{Z} : |x| < 4\}$ in example 6 above. Here $x \in \mathbb{Z}$, so x is a number (not a set), and thus $|x|$ necessarily means absolute value, not cardinality. On the other hand, suppose that $A = \{\{1, 2\}, \{3, 4, 5, 6\}, \{7\}\}$, and $B = \{X \in A : |X| < 3\}$. The elements of A are sets (not numbers), so $|X|$ in the expression for B must mean cardinality. Therefore $B = \{\{1, 2\}, \{7\}\}$.

We'll close this section with a summary of special sets or types of sets that arise so often they're given special names and symbols.

- The empty set: $\varnothing = \{\}$
- The natural numbers: $\mathbb{N} = \{1, 2, 3, \ldots\}$
- The integers: $\mathbb{Z} = \{\ldots, -3, -2, -1, 0, 1, 2, 3, \ldots\}$
- The rational numbers: $\mathbb{Q} = \{x : x = m/n, \text{ where } m, n \in \mathbb{Z} \text{ and } n \neq 0\}$

- The real numbers: $\mathbb{R}$ (the set of all real numbers on the number line)

Note that $\mathbb{R}$ contains both rational and irrational numbers, so $\mathbb{Q} \neq \mathbb{R}$ (for example, $\sqrt{2} \in \mathbb{R}$ but $\sqrt{2} \notin \mathbb{Q}$).

The following special sets form various intervals on the number line, where $a, b \in \mathbb{R}$ with $a < b$.

- Closed interval: $[a, b] = \{x \in \mathbb{R} : a \leq x \leq b\}$
- Half-open interval: $(a, b] = \{x \in \mathbb{R} : a < x \leq b\}$
- Half-open interval: $[a, b) = \{x \in \mathbb{R} : a \leq x < b\}$
- Open interval: $(a, b) = \{x \in \mathbb{R} : a < x < b\}$
- Infinite interval: $(a, \infty) = \{x \in \mathbb{R} : a < x\}$
- Infinite interval: $[a, \infty) = \{x \in \mathbb{R} : a \leq x\}$
- Infinite interval: $(-\infty, b) = \{x \in \mathbb{R} : x < b\}$
- Infinite interval: $(-\infty, b] = \{x \in \mathbb{R} : x \leq b\}$

Keep in mind that these intervals are on the number line, so they have infinitely many elements. The set $(0.1, 0.2)$, for example, contains infinitely many numbers, even though the end points are "close" together. Note another notational conflict: the symbol (a, b) can denote both an interval on the real line as well as a point on a coordinate plane. Fortunately, the difference is always clear from context (or can be made clear). In the next section, we'll see yet another meaning of (a, b).

The Cartesian Product

Given two sets A and B, it's possible to "multiply" them to produce a new set, denoted by $A \times B$. This result is called the cartesian product of A and B. To understand it, we first need the concept of an ordered pair.

Definition An **ordered pair** is a list (x, y) of two entities x and y, enclosed in parentheses and separated by a comma.

$(2, 4)$ is an ordered pair, for example, as is $(4, 2)$. These two ordered pairs differ because, even though they contain the same entities, the order of those entities differs, so $(2, 4) \neq (4, 2)$. Ordered pairs can be used to describe points on the plane, and more. The entities in an ordered pair aren't restricted to only numbers; any comma-separated list of two entities enclosed by parentheses is an ordered pair, for example,

- Ordered pairs of letters, such as (s, t)
- Ordered pairs of sets, such as ({1, 4}, {2, 1})
- Ordered pairs of ordered pairs, such as ((3, 5), (5, 3))
- Ordered pairs of mixed types of entities, such as (2, {1, 2, 3}) and ($\mathbb{R}$, (0, 0))

We define the cartesian product in terms of ordered pairs.

> **Definition** The **cartesian product** of two sets A and B is another set, denoted by $A \times B$ and defined as $A \times B = \{(a, b) : a \in A, b \in B\}$.

Thus $A \times B$ is a set of ordered pairs of elements from A and B. For example, if $A = \{k, l, m\}$ and $B = \{q, r\}$, then

$$A \times B = \{(k, q), (k, r), (l, q), (l, r), (m, q), (m, r)\}$$

The following figure shows how to make a schematic diagram of $A \times B$. Arrange the elements of A horizontally and the elements of B vertically, as if A and B form an x-axis and y-axis. Then fill in the ordered pairs so that each element (x, y) is in the column headed by x and the row headed by y.

B				A × B
r	(k, r)	(l, r)	(m, r)	
q	(k, q)	(l, q)	(m, q)	
	k	l	m	A

As an exercise, draw a similar diagram for $\{0, 1\} \times \{2, 1\} = \{(0, 2), (0, 1), (1, 2), (1, 1)\}$. The rectangular array of such diagrams gives rise to the following general fact about the cardinality of a cartesian product.

> **Fact** If A and B are finite sets, then $|A \times B| = |A| \cdot |B|$.

The familiar set $\mathbb{R} \times \mathbb{R} = \{(x, y) : x, y \in \mathbb{R}\}$ can represent the set of points on the cartesian plane, as shown in the following figure (left). The set $\mathbb{R} \times \mathbb{N} = \{(x, y) : x \in \mathbb{R}, y \in \mathbb{N}\}$ can be regarded as all the points on the cartesian plane whose second coordinate is a natural number, as shown in the following figure (middle), which illustrates that $\mathbb{R} \times \mathbb{N}$ looks like infinitely many horizontal lines at integer heights above the x axis. The set $\mathbb{N} \times \mathbb{N}$ can be visualized as the set of all points on the cartesian plane whose coordinates are both natural numbers. $\{(x, y) : x, y \in \mathbb{N}\}$ looks like a grid of dots in the first quadrant, as shown in following figure (right).

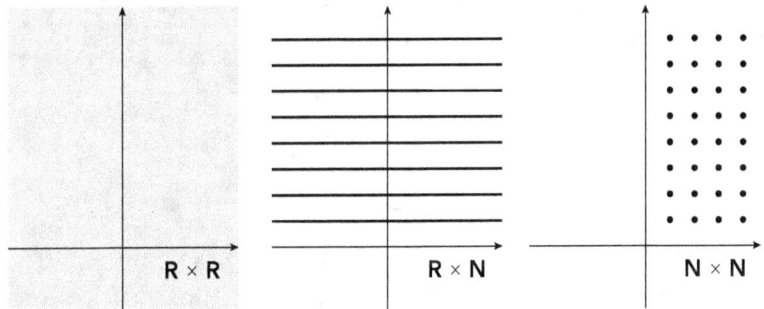

It's possible for one factor of a cartesian product to be a cartesian product itself, as in $\mathbb{R} \times (\mathbb{N} \times \mathbb{Z}) = \{(x, (y, z)) : x \in \mathbb{R}, (y, z) \in \mathbb{N} \times \mathbb{Z}\}$.

We can move beyond ordered pairs to define cartesian products of three or more sets. An **ordered triple** is a list (x, y, z). The cartesian product of the three sets $\mathbb{R}$, $\mathbb{N}$, and $\mathbb{Z}$ is $\mathbb{R} \times \mathbb{N} \times \mathbb{Z} = \{(x, y, z) : x \in \mathbb{R}, y \in \mathbb{N}, z \in \mathbb{Z}\}$. Of course, we needn't stop at ordered triples. In general,

$$A_1 \times A_2 \times \cdots \times A_n = \{(x_1, x_2, \ldots, x_n) : x_i \in A_i \text{ for each } i = 1, 2, \ldots, n\}$$

Be careful with parentheses. The expressions $\mathbb{R} \times (\mathbb{N} \times \mathbb{Z})$ and $\mathbb{R} \times \mathbb{N} \times \mathbb{Z}$ differ. The first expression is a cartesian product of two sets; its elements are ordered pairs $(x, (y, z))$. The second expression is a cartesian product of three sets; its elements are ordered triples (x, y, z). In many situations, there's little ambiguity in blurring the distinction between expressions like $(x, (y, z))$ and (x, y, z), but for this book we consider them to differ.

We can also take **cartesian powers** of sets. For any set A and positive integer n, the power A^n is the cartesian product of A with itself n times:

$$A^n = A \times A \times \cdots \times A = \{(x_1, x_2, \ldots, x_n) : x_1, x_2, \ldots, x_n \in A\}$$

In this way, $\mathbb{R}^2$ is the familiar cartesian plane and $\mathbb{R}^3$ is a three-dimensional space. You can visualize how, if $\mathbb{R}^2$ is the plane, then $\mathbb{Z}^2 = \{(m, n) : m, n \in \mathbb{Z}\}$ is a grid of points on the plane. Likewise, as $\mathbb{R}^3$ is three-dimensional space, $\mathbb{Z}^3 = \{(m, n, p) : m, n, p \in \mathbb{Z}\}$ is a grid of points in space.

In other contexts you might encounter sets that appear to be similar to $\mathbb{R}^n$ but actually differ from it. Consider, for example, the set of all two-by-three matrices with entries from $\mathbb{R}$:

$$M = \left\{ \begin{bmatrix} u & v & w \\ x & y & z \end{bmatrix} : u, v, w, x, y, z \in \mathbb{R} \right\}$$

On the surface, M doesn't much differ from the set

$$\mathbb{R}^6 = \{(u, v, w, x, y, z) : u, v, w, x, y, z \in \mathbb{R}\}$$

because both sets are merely arrangements of six real numbers. Despite their similarity, however, $M \neq \mathbb{R}^6$ because two-by-three matrices aren't the same entities as six-number sequences.

Subsets

It can happen that every element of some set A is also an element of another set B. Each element of $A = \{0, 2, 4\}$ is also an element of $B = \{0, 1, 2, 3, 4\}$, for example. When A and B are related in this way we say that A is a subset of B.

> **Definition** Let A and B be sets. If every element of A is also an element of B, then A is a **subset** of B, denoted by $A \subseteq B$. A is *not* a subset of B, denoted by $A \not\subseteq B$, if it's *not* true that every element of A is also an element of B. In other words, $A \not\subseteq B$ means that there is at least one element of A that is *not* an element of B.

The following statements are all true:

1. $\{2, 3, 7\} \subseteq \{2, 3, 4, 5, 6, 7\}$
2. $\{2, 3, 7\} \not\subseteq \{2, 4, 5, 6, 7\}$

3. $\{2, 3, 7\} \subseteq \{2, 3, 7\}$

4. $\{2n : n \in \mathbb{Z}\} \subseteq \mathbb{Z}$

5. $\{(x, \sin(x)) : x \in \mathbb{R}\} \subseteq \mathbb{R}^2$

6. $\{2, 3, 5, 7, 11, 13, \ldots\} \subseteq \mathbb{N}$

7. $\mathbb{N} \subseteq \mathbb{Z} \subseteq \mathbb{Q} \subseteq \mathbb{R}$

8. $\mathbb{R} \times \mathbb{N} \subseteq \mathbb{R} \times \mathbb{R}$

If B is *any* set, then $\emptyset \subseteq B$. To see why this assertion is true, note that the last sentence of the definition above states that the symbol $\emptyset \subseteq B$ means there's at least one element of $\emptyset$ that's *not* an element of B. But this can't be true because $\emptyset$ contains no elements. Thus it's not the case that $\emptyset \not\subseteq B$, so it must be true that $\emptyset \subseteq B$.

Fact The empty set is a subset of every set; that is, $\emptyset \subseteq B$ for any set B.

This concept of "building" subsets incrementally can be used to list all the subsets of a given set B. Let $B = \{a, b, c\}$, for example. One approach to listing all the subsets of B is to create a tree-like structure. Start with the subset $\{\}$, which is shown on the left of the following figure.

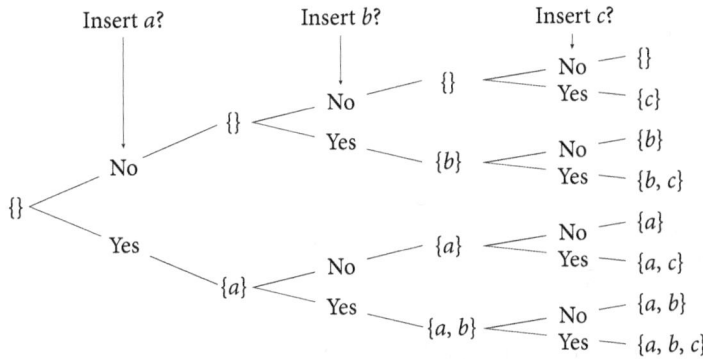

Considering the element a of B, we have a choice: insert a or not. The lines from $\{\}$ point to the results of this choice: either $\{\}$ or $\{a\}$. Next is element b of B. For each of the sets just formed we can either insert or not insert b, and the lines on the diagram point to the resulting sets: $\{\}$, $\{b\}$, $\{a\}$, or $\{a, b\}$. Finally, for each of these sets, we can either insert c or not insert it, yielding the rightmost column: the sets {}, {c}, {b}, {b, c}, {a}, {a, c}, {a, b}, and {a, b, c} are the eight subsets of $B = \{a, b, c\}$.

We can see from the branches of this tree that if, say, $B = \{a\}$, then B would have only two subsets: those in the second column of the diagram. If $B = \{a, b\}$, then B would have four subsets: those listed in the third column, and so on. At each branching of the tree, the number of subsets doubles. Thus in general, if $|B| = n$, then B must have 2^n subsets.

Fact If a finite set has n elements, then it has 2^n subsets.

The following example lists all the subsets of the set $B = \{1, 2, \{1, 3\}\}$, which has three elements: 1, 2, and $\{1, 3\}$. For such a small set, it's unnecessary to draw a tree to list its $2^3 = 8$ subsets:

$$\{\}, \{1\}, \{2\}, \{\{1, 3\}\}, \{1, 2\}, \{1, \{1, 3\}\}, \{2, \{1, 3\}\}, \{1, 2, \{1, 3\}\}$$

Examples like this help you recognize what is and isn't a subset. You know immediately that a set such as $\{1, 3\}$, for example, is *not* a subset of B because it can't be created by selecting elements from B: 3 isn't an element of B and thus it's not a valid selection. Note that although $\{1, 3\} \not\subseteq B$, it's true that $\{1, 3\} \in B$ and $\{\{1, 3\}\} \subseteq B$.

Each of the following statements is true and illustrates an aspect of set theory as we've described it so far.

$1 \in \{1, \{1\}\}$
   1 is the first element of $\{1, \{1\}\}$.

$1 \not\subseteq \{1, \{1\}\}$
   1 isn't a set.

$\{1\} \in \{1, \{1\}\}$
   $\{1\}$ is the second element of $\{1, \{1\}\}$.

$\{1\} \subseteq \{1, \{1\}\}$
   Create the subset $\{1\}$ by selecting 1 from $\{1, \{1\}\}$.

$\{\{1\}\} \notin \{1, \{1\}\}$
   $\{1, \{1\}\}$ contains only 1 and $\{1\}$, and not $\{\{1\}\}$.

$\{\{1\}\} \subseteq \{1, \{1\}\}$
   Create the subset $\{\{1\}\}$ by selecting $\{1\}$ from $\{1, \{1\}\}$.

$\mathbb{N} \notin \mathbb{N}$
> $\mathbb{N}$ is a set (not a number) and $\mathbb{N}$ contains only numbers.

$\mathbb{N} \subseteq \mathbb{N}$
> $X \subseteq X$ for every set X.

$\varnothing \notin \mathbb{N}$
> The set $\mathbb{N}$ contains only numbers and no sets.

$\varnothing \subseteq \mathbb{N}$
> $\varnothing$ is a subset of every set.

$\mathbb{N} \in \{\mathbb{N}\}$
> $\{\mathbb{N}\}$ has only one element, the set $\mathbb{N}$.

$\mathbb{N} \not\subseteq \{\mathbb{N}\}$
> $1 \in \mathbb{N}$, for example, but $1 \notin \{\mathbb{N}\}$.

$\varnothing \notin \{\mathbb{N}\}$
> The only element of $\{\mathbb{N}\}$ is $\mathbb{N}$, and $\mathbb{N} \neq \varnothing$.

$\varnothing \subseteq \{\mathbb{N}\}$
> $\varnothing$ is a subset of every set.

$\varnothing \in \{\varnothing, \mathbb{N}\}$
> $\varnothing$ is the first element of $\{\varnothing, \mathbb{N}\}$.

$\varnothing \subseteq \{\varnothing, \mathbb{N}\}$
> $\varnothing$ is a subset of every set.

$\{\mathbb{N}\} \subseteq \{\varnothing, \mathbb{N}\}$
> Create the subset $\{\mathbb{N}\}$ by selecting $\mathbb{N}$ from $\{\varnothing, \mathbb{N}\}$.

$\{\mathbb{N}\} \not\subseteq \{\varnothing, \{\mathbb{N}\}\}$
> $\mathbb{N} \notin \{\varnothing, \{\mathbb{N}\}\}$.

$\{\mathbb{N}\} \in \{\varnothing, \{\mathbb{N}\}\}$
> $\{\mathbb{N}\}$ is the second element of $\{\varnothing, \{\mathbb{N}\}\}$.

$\{(1, 2), (2, 3), (8, 1)\} \subseteq \mathbb{N} \times \mathbb{N}$
> Each of $(1, 2)$, $(2, 3)$, and $(8, 1)$ is in $\mathbb{N} \times \mathbb{N}$.

The examples above, while useful for illustrating the concept of subsets, are a bit artificial. In general, subsets arise naturally. For example, consider the unit circle $C = \{(x, y) \in \mathbb{R}^2 : x^2 + y^2 = 1\}$. C is a subset of $\mathbb{R}^2$ (that is, $C \subseteq \mathbb{R}^2$). Likewise the graph of a function $y = f(x)$ is a set of points $G = \{(x, f(x)) : x \in \mathbb{R}\}$, and $G \subseteq \mathbb{R}^2$. Sets such as C and G are more easily understood or visualized when regarded as subsets of $\mathbb{R}^2$. Mathematics is filled with examples where it's important to regard one set as a subset of another.

Power Sets

Given a set, you can form a new set with the power set operation, defined as follows.

> **Definition** If A is a set, then the **power set** of A, denoted by $P(A)$, is the set of all subsets of A. That is, $P(A) = \{X : X \subseteq A\}$.

Suppose that $A = \{1, 2, 3\}$. All the subsets of A are $\{\}, \{1\}, \{2\}, \{3\}, \{1, 2\}, \{1, 3\}, \{2, 3\}$ and $\{1, 2, 3\}$. Thus the power set of A is

$$P(A) = \{\emptyset, \{1\}, \{2\}, \{3\}, \{1, 2\}, \{1, 3\}, \{2, 3\}, \{1, 2, 3\}\}$$

As stated in the preceding section, if a finite set A has n elements, then it has 2^n subsets, thus its power set contains 2^n elements.

> **Fact** If A is a finite set, then $|P(A)| = 2^{|A|}$.

The following statements are all *true*:

1. $P(\{0, 1, 3\}) = \{\emptyset, \{0\}, \{1\}, \{3\}, \{0, 1\}, \{0, 3\}, \{1, 3\}, \{0, 1, 3\}\}$
2. $P(\{1, 2\}) = \{\emptyset, \{1\}, \{2\}, \{1, 2\}\}$
3. $P(\{1\}) = \{\emptyset, \{1\}\}$
4. $P(\emptyset) = \{\emptyset\}$
5. $P(\{a\}) = \{\emptyset, \{a\}\}$
6. $P(\{\emptyset\}) = \{\emptyset, \{\emptyset\}\}$
7. $P(\{a\}) \times P(\{\emptyset\}) = \{(\emptyset, \emptyset), (\emptyset, \{\emptyset\}), (\{a\}, \emptyset), (\{a\}, \{\emptyset\})\}$
8. $P(P(\{\emptyset\})) = \{\emptyset, \{\emptyset\}, \{\{\emptyset\}\}, \{\emptyset, \{\emptyset\}\}\}$

9. $P(\{1, \{1, 2\}\}) = \{\varnothing, \{1\}, \{\{1, 2\}\}, \{1, \{1, 2\}\}\}$

10. $P(\{\mathbb{Z}, \mathbb{N}\}) = \{\varnothing, \{\mathbb{Z}\}, \{\mathbb{N}\}, \{\mathbb{Z}, \mathbb{N}\}\}$

The following statements are all *false*:

$P(1) = \{\varnothing, \{1\}\}$
This assertion is meaningless because 1 isn't a set.

$P(\{1, \{1, 2\}\}) = \{\varnothing, \{1\}, \{1, 2\}, \{1, \{1, 2\}\}\}$
This assertion is false because $\{1, 2\} \not\subseteq \{1, \{1, 2\}\}$.

$P(\{1, \{1, 2\}\}) = \{\varnothing, \{\{1\}\}, \{\{1, 2\}\}, \{1, \{1, 2\}\}\}$
This assertion is false because $\{\{1\}\} \not\subseteq \{1, \{1, 2\}\}$.

If A is finite, then it's possible (though perhaps impractical) to list all the elements of $P(A)$ between braces, as in the examples above. Such a listing is impossible if A is infinite. $P(\mathbb{N})$, for example, has infinitely many subsets. In fact, it's even unclear how the subsets could be listed in a sensible pattern:

$$P(\mathbb{N}) = \{\varnothing, \{1\}, \{2\}, \ldots, \{1, 2\}, \{1, 3\}, \ldots, \{28, 36\}, \ldots,$$
$$\{3, 54, 152\}, \ldots, \{2, 4, 6, 8, \ldots\}, \ldots, ?, \ldots\}$$

The set $P(\mathbb{R}^2)$ is incomprehensibly large. Think of $\mathbb{R}^2 = \{(x, y) : x, y \in \mathbb{R}\}$ as the set of all points on the cartesian plane. A subset of $\mathbb{R}^2$ (that is, an *element* of $P(\mathbb{R}^2)$) is a set of points in the plane. Let's look at a few of these sets. Because $\{(0, 0), (1, 1)\} \subseteq \mathbb{R}^2$, we know that $\{(0, 0), (1, 1)\} \in P(\mathbb{R}^2)$. An illustration of this subset is shown in the following figure (left).

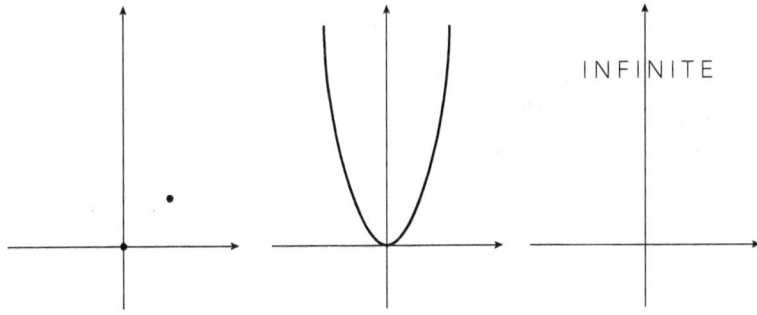

Another example: the graph of the equation $y = x^2$ is the set of points $G = \{(x, x^2) : x \in \mathbb{R}\}$ and this is a subset of $\mathbb{R}^2$, so $G \in P(\mathbb{R}^2)$. G is shown in the preceding figure (middle). Extending this idea, we can draw the graph of any function f where $f: \mathbb{R} \to \mathbb{R}$ is an element of $P(\mathbb{R}^2)$. In fact, any black-and-white image on the plane can be thought of as a subset of $\mathbb{R}^2$, where the black points belong to the subset and the white points do not. So the text "INFINITE" in the preceding figure (right) is a subset of $\mathbb{R}^2$ and therefore an element of $P(\mathbb{R}^2)$. By that token, $P(\mathbb{R}^2)$ contains a copy of the page you're now reading.

Thus in addition to containing every imaginable function and every imaginable black-and-white image, $P(\mathbb{R}^2)$ also contains the full text of every book that has ever been written, that will be written, and that will never be written, including your biography. For a truly mind-boggling set, consider $P(P(\mathbb{R}^2))$.

Union, Intersection, and Difference

Just as various operations such as addition, subtraction, and multiplication can be applied to numbers, so can various operations be applied to sets. The cartesian product (page 5) is one such operation: given sets A and B, we can combine them by using the $\times$ operator to get a new set $A \times B$. Here are three more set operations:

Definition Suppose that A and B are sets.

The **union** of A and B is the set $A \cup B = \{x : x \in A \text{ or } x \in B\}$.

The **intersection** of A and B is the set $A \cap B = \{x : x \in A \text{ and } x \in B\}$.

The **difference** of A and B is the set $A - B = \{x : x \in A \text{ and } x \notin B\}$.

In words, the union $A \cup B$ is the set of all elements that are in A or in B (or in both). The intersection $A \cap B$ is the set of all elements in both A and B. The difference $A - B$ is the set of all elements that are in A but not in B.

In the following examples, $A = \{a, b, c, d, e\}$, $B = \{d, e, f\}$, and $C = \{1, 2, 3\}$.

1. $A \cup B = \{a, b, c, d, e, f\}$
2. $A \cap B = \{d, e\}$

3. $A - B = \{a, b, c\}$
4. $B - A = \{f\}$
5. $(A - B) \cup (B - A) = \{a, b, c, f\}$
6. $A \cup C = \{a, b, c, d, e, 1, 2, 3\}$
7. $A \cap C = \emptyset$
8. $A - C = \{a, b, c, d, e\}$
9. $(A \cap C) \cup (A - C) = \{a, b, c, d, e\}$
10. $(A \cap B) \times B = \{(d, d), (d, e), (d, f), (e, d), (e, e), (e, f)\}$
11. $(A \times C) \cap (B \times C) = \{(d, 1), (d, 2), (d, 3), (e, 1), (e, 2), (e, 3)\}$

Observe that for any sets X and Y it is always true that $X \cup Y = Y \cup X$ and $X \cap Y = Y \cap X$, but in general $X - Y \neq Y - X$.

The following examples use interval notation, so, for example, $[2, 5] = \{x \in \mathbb{R} : 2 \leq x \leq 5\}$. Sketching these examples on the number line will help you visualize them.

1. $[2, 5] \cup [3, 6] = [2, 6]$
2. $[2, 5] \cap [3, 6] = [3, 5]$
3. $[2, 5] - [3, 6] = [2, 3)$
4. $[0, 3] - [1, 2] = [0, 1) \cup (2, 3]$

Next, let $A = \{(x, x^2) : x \in \mathbb{R}\}$ be the graph of the equation $y = x^2$ and let $B = \{(x, x + 2) : x \in \mathbb{R}\}$ be the graph of the equation $y = x + 2$. These sets are subsets of $\mathbb{R}^2$ and are both sketched in the following figure.

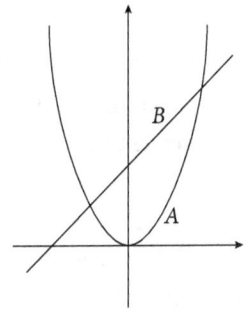

The following figure shows $A \cup B$, $A \cap B$, and $A - B$. $A \cup B = \{(x, y) : x \in \mathbb{R}, y = x^2 \text{ or } y = x + 2\}$ is the set of all points (x, y) that are on one (or both) of the two graphs. $A \cap B = \{(-1, 1), (2, 4)\}$ consists of only two elements, the two points where the graphs intersect. $A - B = \{(x, x^2) : x \in \mathbb{R} - \{-1, 2\}\}$ is the set A with "holes" where B crossed it.

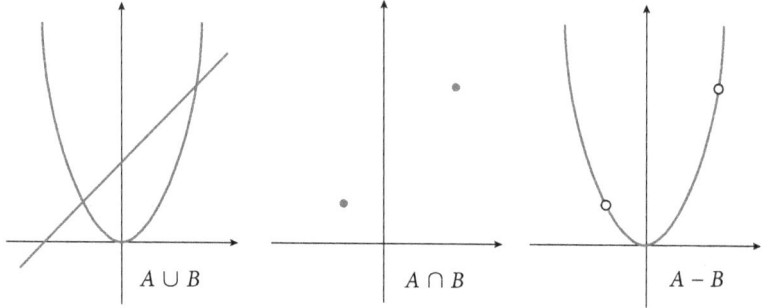

Complement

This section introduces the set operation called the set complement. To understand it, we first need the concept of a universal set.

When working with a set, we almost always regard it as a subset of some larger set. For example, consider the set of prime numbers

$$P = \{2, 3, 5, 7, 11, 13, \ldots\}$$

If asked to name some entities that aren't in P, we might mention some composite numbers such as 4 or 100 or 435. It probably wouldn't occur to us to say that Brad Pitt is not in P. True, Brad Pitt isn't in P, but he lies entirely outside of the discussion of which numbers are prime. We have an unstated assumption that

$$P \subseteq \mathbb{N}$$

because $\mathbb{N}$ is the most natural setting in which to discuss prime numbers. In this context, anything not in P is still in $\mathbb{N}$. This larger set $\mathbb{N}$ is called the **universal set** or **universe** for P.

Almost every useful set in mathematics can be regarded as having some natural universal set. For example, the unit circle is the set $C = \{(x, y) \in \mathbb{R}^2 : x^2 + y^2 = 1\}$, and because all these points are in the plane $\mathbb{R}^2$ it's natural to regard $\mathbb{R}^2$ as the universal set for C. In the absence of specifics, if A is a set, then its universal set is often denoted by U. We define the complement operation in terms of a universal set.

Definition   Let A be a set with a universal set U. The **complement** of A, denoted by $\overline{A}$, is the set $\overline{A} = U - A$.

Let's look at some examples. If P is the set of prime numbers, then

$$\overline{P} = \mathbb{N} - P = \{1, 4, 6, 8, 9, 10, 12, \ldots\}$$

Thus $\overline{P}$ is the set of composite numbers and 1.

Next, let $A = \{(x, x^2) : x \in \mathbb{R}\}$ be the graph of the equation $y = x^2$. The following figure (left) shows A in its universal set $\mathbb{R}^2$. The complement of A is $\overline{A} = \mathbb{R}^2 - A = \{(x, y) \in \mathbb{R}^2 : y \neq x^2\}$, as shown by the shaded area in the following figure (right).

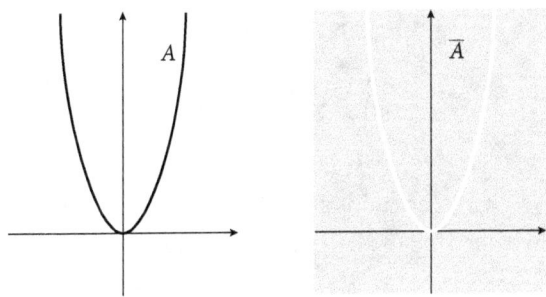

Venn Diagrams

It's sometimes helpful to draw informal, schematic diagrams of sets. In doing so we often represent a set with a circle (or oval), which we regard as enclosing all the elements of the set. Such diagrams can illustrate how sets combine by using various operations. For example, the following figure shows two sets A and B that overlap in a middle region. The sets $A \cup B$, $A \cap B$, and $A - B$ are shaded.

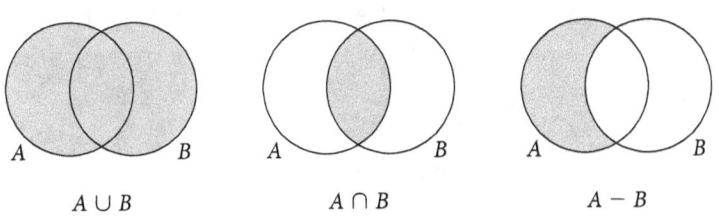

Chapter 1 Sets   17

Such graphical representations of sets are called **Venn diagrams**, after their inventor, British logician John Venn, 1834–1923.

Though you're unlikely to draw Venn diagrams as a part of a proof of any theorem, you can use them for "scratch work" to understand how sets combine, develop strategies for proving certain theorems, or solve certain problems. The remainder of this section uses Venn diagrams to explore how three sets can be combined by using ∪ and ∩.

Let's start with the set $A \cup B \cup C$. This set contains all elements that are in one or more of the sets A, B, and C. The following figure shows a Venn diagram for this set.

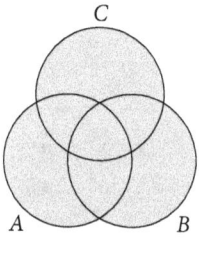

$A \cup B \cup C$

The set $A \cap B \cap C$ contains all elements common to each of A, B, and C. The region belonging to all three sets is shaded in the following figure.

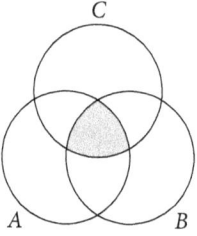

$A \cap B \cap C$

We can also think of $A \cap B \cap C$ as the two-step operation $(A \cap B) \cap C$. In this expression the set $A \cap B$ is represented by the region common to both A and B, and when *this* region intersects with C we get the result shown in the preceding figure. This exercise is a visual representation of the fact that $A \cap B \cap C = (A \cap B) \cap C$. Similarly, we have $A \cap B \cap C = A \cap (B \cap C)$. Likewise, $A \cup B \cup C = (A \cup B) \cup C = A \cup (B \cup C)$.

Notice that in the preceding examples, where each expression contains either only the symbol ∪ or only the symbol ∩, the placement of parentheses is irrelevant, so we're free to omit them, which is analogous to the situation in algebra involving the expressions $(a + b) + c = a + (b + c)$ or $(a \cdot b) \cdot c = a \cdot (b \cdot c)$. We tend to omit the parentheses and write simply $a + b + c$ or $a \cdot b \cdot c$. By contrast, in an algebraic expression like $(a + b) \cdot c$ the parentheses are essential because $(a + b) \cdot c$ and $a + (b \cdot c)$ are generally not equal.

Let's use Venn diagrams to understand the expressions $(A \cup B) \cap C$ and $A \cup (B \cap C)$, which use a mix of ∪ and ∩ operators. The following figure shows a Venn diagram for $(A \cup B) \cap C$. In the drawing on the left, the set $A \cup B$ is hatched horizontally, while C is hatched vertically. Thus the set $(A \cup B) \cap C$ is represented by the cross-hatched region where $A \cup B$ and C overlap. The superfluous shadings are omitted in the drawing on the right showing the set $(A \cup B) \cap C$.

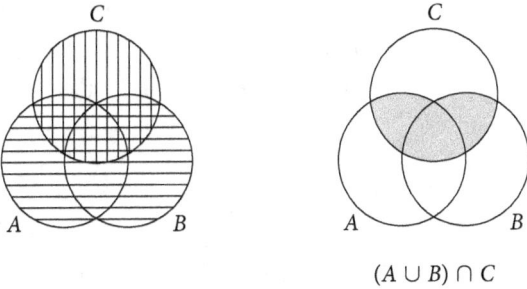

$(A \cup B) \cap C$

Let's repeat this exercise for $A \cup (B \cap C)$. In the following figure the set A is hatched horizontally, and $B \cap C$ is hatched vertically. The union $A \cup (B \cap C)$ is represented by the totality of all shaded regions, as shown on the right.

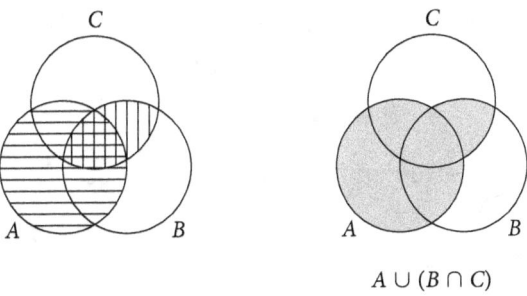

$A \cup (B \cap C)$

Compare the two preceding figures for $(A \cup B) \cap C$ and $A \cup (B \cap C)$. The fact that the diagrams are different indicates that $(A \cup B) \cap C \neq A \cup (B \cap C)$ in general. Thus an expression such as $A \cup B \cap C$ is meaningless because it might mean either $(A \cup B) \cap C$ or $A \cup (B \cap C)$. In summary, Venn diagrams have helped us understand why parentheses are essential in expressions that use both $\cup$ and $\cap$.

Indexed Sets

When a mathematical problem involves many sets, it's often convenient to represent them by using subscripts (also called indices). Thus instead of denoting three sets as A, B, and C, we might instead write them as A_1, A_2, and A_3. These sets are called **indexed sets**.

We defined union and intersection to be operations that combine two sets, but we can apply these operations repeatedly to combine three or more sets. Venn diagrams for the union and intersection of three sets were illustrated in the preceding section, for example. We can now define these concepts rigorously. Given sets $A_1, A_2, A_3, \ldots, A_n$, the set $A_1 \cup A_2 \cup A_3 \cup \cdots \cup A_n$ consists of everything that's in *at least one* of the sets A_i. Likewise $A_1 \cap A_2 \cap A_3 \cap \cdots \cap A_n$ consists of everything that's common to *all* the sets A_i.

Definition Suppose that $A_1, A_2, A_3, \ldots, A_n$ are sets. Then

$A_1 \cup A_2 \cup A_3 \cup \cdots \cup A_n = \{x : x \in A_i$ for *at least one set* A_i, for $1 \leq i \leq n\}$

$A_1 \cap A_2 \cap A_3 \cap \cdots \cap A_n = \{x : x \in A_i$ for *every set* A_i, for $1 \leq i \leq n\}$

These expressions are cumbersome when the number of sets n is large, so we'll use a notational shortcut akin to sigma (Σ) notation. Sigma notation provides convenient symbols for summing many numbers. Given the numbers $a_1, a_2, a_3, \ldots, a_n$, for example, then

$$\sum_{i=1}^{n} a_i = a_1 + a_2 + a_3 + \cdots + a_n$$

The sum is often meaningful even for an infinite list of numbers:

$$\sum_{i=1}^{\infty} a_i = a_1 + a_2 + a_3 + \cdots + a_i + \cdots$$

The notation for the union and intersection operators is similar to sigma notation. Given sets $A_1, A_2, A_3, \ldots, A_n$, we define

$$\bigcup_{i=1}^{n} A_i = A_1 \cup A_2 \cup A_3 \cup \cdots \cup A_n$$

and

$$\bigcap_{i=1}^{n} A_i = A_1 \cap A_2 \cap A_3 \cap \cdots \cap A_n$$

For example, if $A_1 = \{0, 2, 5\}$, $A_2 = \{1, 2, 5\}$, and $A_3 = \{2, 5, 7\}$, then

$$\bigcup_{i=1}^{3} A_i = A_1 \cup A_2 \cup A_3 = \{0, 1, 2, 5, 7\}$$

and

$$\bigcap_{i=1}^{3} A_i = A_1 \cap A_2 \cap A_3 = \{2, 5\}$$

This notation is also used when the list of sets $A_1, A_2, A_3, \ldots$ is infinite:

$$\bigcup_{i=1}^{\infty} A_i = A_1 \cup A_2 \cup A_3 \cup \cdots$$
$$= \{x : x \in A_i \text{ for at least one set } A_i \text{ with } 1 \leq i\}$$

and

$$\bigcap_{i=1}^{\infty} A_i = A_1 \cap A_2 \cap A_3 \cap \cdots$$
$$= \{x : x \in A_i \text{ for every set } A_i \text{ with } 1 \leq i\}$$

For example, if we have the infinite list of sets

$$A_1 = \{-1, 0, 1\}, A_2 = \{-2, 0, 2\}, A_3 = \{-3, 0, 3\}, \ldots, A_i = \{-i, 0, i\}, \ldots$$

then $\bigcup_{i=1}^{\infty} A_i = \mathbb{Z}$ and $\bigcap_{i=1}^{\infty} A_i = \{0\}$.

We can also write

$$\bigcup_{i=1}^{3} A_i = \bigcup_{i \in \{1,2,3\}} A_i$$

to take the union of the sets A_i for $i = 1, 2, 3$.

Likewise:

$$\bigcap_{i=1}^{3} A_i = \bigcap_{i \in \{1,2,3\}} A_i$$

$$\bigcup_{i=1}^{\infty} A_i = \bigcup_{i \in \mathbb{N}} A_i$$

$$\bigcap_{i=1}^{\infty} A_i = \bigcap_{i \in \mathbb{N}} A_i$$

In the preceding equations we're taking the union or intersection of a collection of sets A_i where i is an element of some set, be it $\{1, 2, 3\}$ or $\mathbb{N}$. In general, we have a collection of sets A_i for $i \in I$, where I is the set of possible subscripts. The set I is an **index set**.

The set I needn't consist of integers—we can also subscript by using, say, letters or real numbers. Because we tend to think of i as an integer, we'll use α instead of i to stand for an element of I. Thus we have a collection of sets A_α for $\alpha \in I$, leading to the following definition.

Definition If we have a set A_α for every α in some index set I, then

$$\bigcup_{\alpha \in I} A_\alpha = \{x : x \in A_\alpha \text{ for at least one set } A_\alpha \text{ with } \alpha \in I\}$$

$$\bigcap_{\alpha \in I} A_\alpha = \{x : x \in A_\alpha \text{ for every set } A_\alpha \text{ with } \alpha \in I\}$$

Example 1.1 In this example the sets A_α are subsets of $\mathbb{R}^2$. Let $I = [0, 2] = \{x \in \mathbb{R} : 0 \le x \le 2\}$. For each number $\alpha \in I$, let $A_\alpha = \{(x, \alpha) : x \in \mathbb{R}, 1 \le x \le 2\}$. For instance, given $\alpha = 1 \in I$ the set $A_1 = \{(x, 1) : x \in \mathbb{R}, 1 \le x \le 2\}$ is a horizontal line segment one unit above the x-axis and running between $x = 1$ and $x = 2$, as shown in the following figure (left). Likewise $A_{\sqrt{2}} = \{(x, \sqrt{2}) : x \in \mathbb{R}, 1 \le x \le 2\}$ is a horizontal line segment $\sqrt{2}$ units above the x-axis and running between $x = 1$ and $x = 2$. A few of the other sets A_α are also shown in the figure (left), but they can't all be drawn because there's one A_α for each of the infinitely many numbers $\alpha \in [0, 2]$. The totality of them covers the shaded region in the figure (right), so this region is the union of all the sets A_α. Because the shaded region is the set $\{(x, y) \in \mathbb{R}^2 : 1 \leq x \leq 2, 0 \leq y \leq 2\} = [1, 2] \times [0, 2]$, it follows that

$$\bigcup_{\alpha \in [0,2]} A_\alpha = [1,2] \times [0,2]$$

Likewise, because there's no point (x, y) that's in every set A_α, we have

$$\bigcap_{\alpha \in [0,2]} A_\alpha = \varnothing$$

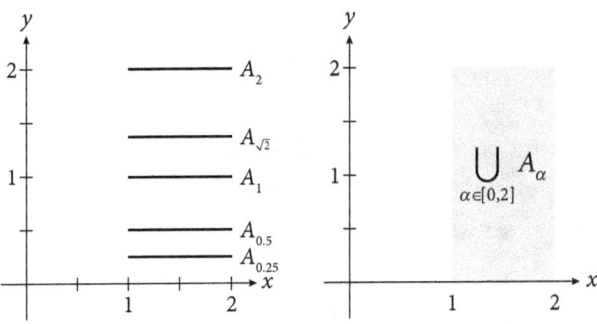

Finally, note that $A_\alpha = [1, 2] \times \{\alpha\}$, so the above expressions can be written as

$$\bigcup_{\alpha \in [0,2]} [1,2] \times \{\alpha\} = [1,2] \times [0,2]$$

and

$$\bigcap_{\alpha \in [0,2]} [1,2] \times \{\alpha\} = \varnothing$$

Example 1.2 In this example the sets are indexed by $\mathbb{R}^2$. For any $(a, b) \in \mathbb{R}^2$, let $P_{(a, b)}$ be the following subset of $\mathbb{R}^3$:

$$P_{(a,b)} = \{(x, y, z) \in \mathbb{R}^3 : ax + by = 0\}$$

In words, given a point $(a, b) \in \mathbb{R}^2$, the corresponding set $P_{(a, b)}$ consists of all points (x, y, z) in $\mathbb{R}^3$ that satisfy the equation $ax + by = 0$. Hence $P_{(a, b)}$ is a plane in $\mathbb{R}^3$, and because every point $(0, 0, z)$ on the z-axis automatically satisfies $ax + by = 0$, every $P_{(a, b)}$ contains the z-axis.

The following figure shows the set $P_{(1,2)} = \{(x, y, z) \in \mathbb{R}^3 : x + 2y = 0\}$. It's the vertical plane that intersects the xy-plane at the line $x + 2y = 0$.

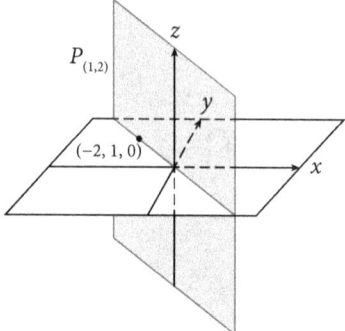

For any point $(a, b) \in \mathbb{R}^2$ with $(a, b) \neq (0, 0)$, we can visualize $P_{(a, b)}$ as the vertical plane that cuts the xy-plane at the line $ax + by = 0$. The following figure shows a few of the $P_{(a, b)}$ planes.

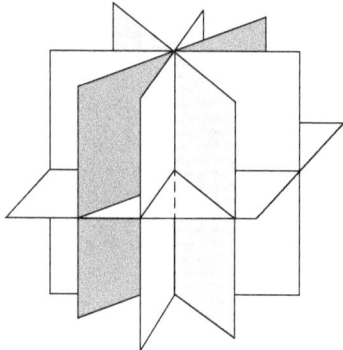

Because any two such planes intersect along the z-axis and the z-axis is a subset of every $P_{(a, b)}$, it's clear that

$$\bigcap_{(a,b) \in \mathbb{R}^2} P_{(a, b)} = \{(0, 0, z) : z \in \mathbb{R}\} = \text{``the } z\text{-axis''}.$$

For the union, note that any given point $(a, b, c) \in \mathbb{R}^3$ belongs to the set $P_{(-b, a)}$ because $(x, y, z) = (a, b, c)$ satisfies the equation $-bx + ay = 0$. (In fact, any (a, b, c) belongs to the special set $P_{(0, 0)} = \mathbb{R}^3$, which is the only $P_{(a, b)}$ that's not a plane.) Because any point in $\mathbb{R}^3$ belongs to some $P_{(a, b)}$ we have

$$\bigcup_{(a,b) \in \mathbb{R}^2} P(a, b) = \mathbb{R}^3$$

The Well-Ordering Principle

In practice, the most interesting sets usually have special properties and structures. The sets $\mathbb{Z}$, $\mathbb{Q}$, and $\mathbb{R}$, for example, are familiar number systems: given such a set, any two of its elements can be added or multiplied to produce another element in the set. These operations obey the familiar commutative, associative, and distributive properties. Such properties lead to the standard algebraic techniques for solving equations. For our purposes, it's unnecessary to define, prove, or verify such properties and techniques; instead, we accept them as fact and base further deductions on them.

We also accept as fact the natural ordering of the elements of $\mathbb{N}$, $\mathbb{Z}$, $\mathbb{Q}$, and $\mathbb{R}$, so that the meaning of "$2 < 4$", for example, is understood and needn't be justified or explained. Similarly, if $x \leq y$ and $a \neq 0$, then we know that $ax \leq ay$ or $ax \geq ay$, depending on whether a is positive or negative.

Another ingrained assumption about the ordering of numbers tells us that any non-empty subset of $\mathbb{N}$ has a smallest element. In other words, if $A \subseteq \mathbb{N}$ and $A \neq \emptyset$, then there exists an element $x_0 \in A$ that's smaller than every other element of A. (To find it, start at 1, then move in increments to 2, 3, 4, and so on, until you arrive at a number $x_0 \in A$; this number is the smallest element of A.) Similarly, given an integer b, any non-empty subset $A \subseteq \{b, b+1, b+2, b+3, \ldots\}$ has a smallest element. This fact is called the **well-ordering principle**. This simple, intuitive idea is often used in proofs.

The well-ordering principle looks simple, but it actually makes a profound and fundamental statement about the positive integers $\mathbb{N}$. In fact, the corresponding statement about the positive real numbers is false: the subset $A = \{1/n : n \in \mathbb{N}\}$ of the positive reals has no smallest element because for any $x_0 = 1/n \in A$ that we choose, there's always a smaller element $1/(n+1) \in A$.

One consequence of the well-ordering principle is the familiar fact that any integer a can be divided by a nonzero integer b, resulting in a quotient q and remainder r. For example, $b = 3$ goes into $a = 17$ $q = 5$ times with remainder $r = 2$. In symbols, $17 = 5 \cdot 3 + 2$, or $a = qb + r$. This fact is called the division algorithm.

Fact (**The Division Algorithm**) Given integers a and b with $b > 0$, there exist integers q and r for which $a = qb + r$ and $0 \le r < b$.

Although we'll accept the division algorithm without proof, note that it follows from the well-ordering principle: given integers a, b with $b > 0$, form the set

$$A = \{a - xb : x \in \mathbb{Z}, 0 \le a - xb\} \subseteq \{0, 1, 2, 3, \ldots\}$$

(For example, if $a = 17$ and $b = 3$ then $A = \{2, 5, 8, 11, 14, 17, 20, \ldots\}$ is the set of positive integers obtained by adding multiples of 3 to 17. Note that the remainder $r = 2$ of $17 \div 3$ is the smallest element of this set.) In general, let r be the smallest element of the set $A = \{a - xb : x \in \mathbb{Z}, 0 \le a - xb\}$. Then $r = a - qb$ for some $x = q \in \mathbb{Z}$, so $a = qb + r$. Moreover, $0 \le r < b$, as follows. The fact that $r \in A \subseteq \{0, 1, 2, 3, \ldots\}$ implies $0 \le r$. In addition, it can't happen that $r \ge b$: if this were true, then the non-negative number $r - b = (a - qb) - b = a - (q+1)b$ having form $a - xb$ would be a smaller element of A than r, and r was explicitly chosen as the smallest element of A. Because it's not true that $r \ge b$, it must be true that $r < b$. Therefore $0 \le r < b$. We've now produced a q and an r for which $a = qb + r$ and $0 \le r < b$.

Moving on, a small issue must be clarified. This book asserts that all of mathematics can be described with sets. But at the same time we've maintained that some mathematical entities aren't sets. For example, our approach states that an individual number, such as 3, isn't itself a set, though it can be an *element* of a set.

We've made this distinction because we need a place to stand as we explore sets. After all, it would appear suspiciously circular to declare that every mathematical entity is a set, and then proceed to define a set as a collection whose members are sets.

But to most mathematicians, saying "The number 3 isn't a set" is like saying "The number 3 isn't a number".

In fact, any number *can* itself be understood as a set. One way to do so is to start with the identification $0 = \varnothing$. Then $1 = \{\varnothing\} = \{0\}$, and $2 = \{\varnothing, \{\varnothing\}\} = \{0, 1\}$, and $3 = \{\varnothing, \{\varnothing\}, \{\varnothing, \{\varnothing\}\}\} = \{0, 1, 2\}$. In general the natural number n is the set $n = \{0, 1, 2, \ldots, n - 1\}$ of the n numbers (which are themselves sets) that come before it.

We won't undertake such a study here, but the elements of the number systems $\mathbb{Z}$, $\mathbb{Q}$, and $\mathbb{R}$ can all be defined in terms of sets. (Even the operations of addition, multiplication, and so on can be defined in set-theoretic terms.) In fact, mathematics itself can be regarded as the study of things that can be described as sets. Any mathematical entity is a set, whether or not we choose to think of it that way.

Russell's Paradox

This section contains optional background material that's not often used in the study of sets, but is interesting nonetheless.

The philosopher and mathematician Bertrand Russell (1872–1970) did groundbreaking work on the theory of sets and the foundations of mathematics. He was probably among the first to understand how the misuse of sets can lead to paradoxical situations. His best-known idea has come to be known as Russell's paradox.

Russell's paradox involves the following set of sets:

$$A = \{X : X \text{ is a set and } X \notin X\} \qquad (1)$$

In words, A is the set of all sets that don't include themselves as elements. Most sets that we can think of are in A. The set $\mathbb{Z}$ of integers isn't an integer (that is, $\mathbb{Z} \notin \mathbb{Z}$) and therefore $\mathbb{Z} \in A$. Also $\emptyset \in A$ because $\emptyset$ is a set and $\emptyset \notin \emptyset$.

Is there a set that's not in A? Consider $B = \{\{\{\ldots\}\}\}$. Think of B as a box containing a box, containing a box, containing a box, and so on, forever. Or a set of Russian dolls, nested one inside the other, endlessly. Curiously, B has only one element, namely B itself:

$$B = \{\underbrace{\{\{\{\ldots\}\}\}}_{B}\}$$

Thus $B \in B$. B doesn't satisfy $B \notin B$, so $B \notin A$ according to equation (1).

Russell's paradox arises from the question, "Is A an element of A?"

For a set X, equation (1) says $X \in A$ means the same thing as $X \notin X$. So for $X = A$, the previous line says $A \in A$ means the same thing as $A \notin A$. The conclusion is Russell's paradox: if $A \in A$ is true, then it's false; if $A \in A$ is false, then it's true.

Initially Russell's paradox sparked a crisis among mathematicians. How could a mathematical statement be both true and false? This notion opposed the very essence of mathematics.

What followed was a careful examination of set theory and an evaluation of what can and can't be regarded as a set. Eventually mathematicians settled upon a collection of axioms for set theory—the **Zermelo–Fraenkel axioms**. One of these axioms is the well-ordering principle. Another, the axiom of foundation, states that no nonempty set X can have the property $X \cap x \neq \emptyset$ for all its elements x, ruling out such circularly defined "sets" as $X = \{X\}$ introduced above. If we adhere to these axioms, then situations like Russell's paradox disappear. Most mathematicians accept all this on faith and ignore the Zermelo–Fraenkel axioms. Paradoxes like Russell's rarely arise in everyday mathematics—you must strive to construct them. Still, Russell's paradox reminds us that precision of thought and language is a crucial part of mathematics.

Problems

1. Write each of the following sets by listing their elements between braces.
   (a) $\{5x - 1 : x \in \mathbb{Z}\}$
   (b) $\{x \in \mathbb{Z} : -2 \leq x < 7\}$
   (c) $\{x \in \mathbb{R} : x^2 = 3\}$
   (d) $\{x \in \mathbb{R} : x^2 + 5x = -6\}$
   (e) $\{x \in \mathbb{R} : \sin \pi x = 0\}$
   (f) $\{x \in \mathbb{Z} : |x| < 5\}$
   (g) $\{x \in \mathbb{Z} : |6x| < 5\}$
   (h) $\{5a + 2b : a, b \in \mathbb{Z}\}$

2. Write each of the following sets in set-builder notation.
   (a) $\{2, 4, 8, 16, 32, 64, \ldots\}$
   (b) $\{\ldots, -9, -6, -3, 0, 3, 6, 9, \ldots\}$
   (c) $\{0, 1, 4, 9, 16, 25, 36, \ldots\}$
   (d) $\{3, 4, 5, 6, 7, 8\}$
   (e) $\{\ldots, 1/8, 1/4, 1/2, 1, 2, 4, 8, \ldots\}$
   (f) $\{\ldots, -\pi, -\pi/2, 0, \pi/2, \pi, 3\pi/2, 2\pi, 5\pi/2, \ldots\}$

3. Find the following cardinalities.
   (a) $|\{\{1\}, \{2, \{3, 4\}\}, \emptyset\}|$
   (b) $|\{\{\{1\}, \{2, \{3, 4\}\}, \emptyset\}\}|$
   (c) $|\{x \in \mathbb{Z} : |x| < 10\}|$
   (d) $|\{x \in \mathbb{Z} : x^2 < 10\}|$
   (e) $|\{x \in \mathbb{N} : x^2 < 0\}|$

4. Sketch the following sets of points in the x–y plane.
   (a) $\{(x, y) : x \in [1, 2], y \in [1, 2]\}$
   (b) $\{(x, y) : x \in [-1, 1], y = 1\}$
   (c) $\{(x, y) : |x| = 2, y \in [0, 1]\}$
   (d) $\{(x, y) : x, y \in \mathbb{R}, x^2 + y^2 = 1\}$
   (e) $\{(x, y) : x, y \in \mathbb{R}, y \geq x^2 - 1\}$
   (f) $\{(x, x + y) : x \in \mathbb{R}, y \in \mathbb{Z}\}$
   (g) $\{(x, y) \in \mathbb{R}^2 : (y - x)(y + x) = 0\}$

5. Write the following sets by listing their elements between braces, where $A = \{1, 2, 3, 4\}$ and $B = \{a, c\}$.
   (a) $A \times B$
   (b) $B \times A$
   (c) $A \times A$
   (d) $B \times B$
   (e) $\emptyset \times B$
   (f) $(A \times B) \times B$
   (g) $A \times (B \times B)$
   (h) B^3

6. Write the following sets by listing their elements between braces.
   (a) $\{x \in \mathbb{R} : x^2 = 2\} \times \{a, c, e\}$
   (b) $\{x \in \mathbb{R} : x^2 = 2\} \times \{x \in \mathbb{R} : |x| = 2\}$
   (c) $\{\emptyset\} \times \{0, \emptyset\} \times \{0, 1\}$

7. Sketch the following cartesian products on the x–y plane $\mathbb{R}^2$ (or $\mathbb{R}^3$ for part (f)).
   (a) $\{1, 2, 3\} \times \{-1, 0, 1\}$
   (b) $[0, 1] \times [0, 1]$
   (c) $\{1, 1\frac{1}{2}, 2\} \times [1, 2]$

Chapter 1 Sets 29

(d) $\{1\} \times [0, 1]$
(e) $\mathbb{N} \times \mathbb{Z}$
(f) $[0, 1] \times [0, 1] \times [0, 1]$

8. List all the subsets of the following sets.
    (a) $\{1, 2, 3, 4\}$
    (b) $\{\{\mathbb{R}\}\}$
    (c) $\{\emptyset\}$
    (d) $\{\mathbb{R}, \{\mathbb{Q}, \mathbb{N}\}\}$

9. Write the following sets by listing their elements between braces.
    (a) $\{X : X \subseteq \{3, 2, a\} \text{ and } |X| = 2\}$
    (b) $\{X : X \subseteq \{3, 2, a\} \text{ and } |X| = 4\}$

10. Decide whether the following statements are true or false.
    (a) $\mathbb{R}^3 \subseteq \mathbb{R}^3$
    (b) $\mathbb{R}^2 \subseteq \mathbb{R}^3$
    (c) $\{(x, y) : x - 1 = 0\} \subseteq \{(x, y) : x^2 - x = 0\}$
    (d) $\{(x, y) : x^2 - x = 0\} \subseteq \{(x, y) : x - 1 = 0\}$

11. Find the indicated sets.
    (a) $P(\{\{a, b\}, \{c\}\})$
    (b) $P(\{\{\emptyset\}, 5\})$
    (c) $P(P(2))$
    (d) $P(\{a, b\}) \times P(\{0, 1\})$
    (e) $P(\{a, b\} \times \{0\})$
    (f) $\{X \subseteq P(\{1, 2, 3\}) : |X| \leq 1\}$

12. Suppose that $|A| = m$ and $|B| = n$. Find the following cardinalities.
    (a) $|P(P(P(A)))|$
    (b) $|P(A \times B)|$
    (c) $|X \in P(A) : |X| \leq 1|$
    (d) $|P(P(P(A \times \emptyset)))|$

13. Let $A = \{4, 3, 6, 7, 1, 9\}$, $B = \{5, 6, 8, 4\}$, and $C = \{5, 8, 4\}$. Find:
    (a) $A \cup B$
    (b) $A \cap B$
    (c) $A - B$
    (d) $A - C$
    (e) $B - A$
    (f) $A \cap C$
    (g) $B \cap C$
    (h) $B \cup C$
    (i) $C - B$

14. Let $A = \{0, 1\}$ and $B = \{1, 2\}$. Find:
    (a) $(A \times B) \cap (B \times B)$
    (b) $(A \times B) \cup (B \times B)$
    (c) $(A \times B) - (B \times B)$
    (d) $(A \cap B) \times A$
    (e) $(A \times B) \cap B$
    (f) $P(A) \cap P(B)$
    (g) $P(A) - P(B)$
    (h) $P(A \cap B)$
    (i) $P(A \times B)$

15. Sketch the sets $X = [1, 3] \times [1, 3]$ and $Y = [2, 4] \times [2, 4]$ on the plane $\mathbb{R}^2$. Also shade in separate sketches the sets $X \cup Y$, $X \cap Y$, $X - Y$, and $Y - X$. (Note that X and Y are cartesian products of intervals, which you also sketched in earlier problems.)

16. Sketch the sets $X = \{(x, y) \in \mathbb{R}^2 : x^2 + y^2 \leq 1\}$ and $Y = \{(x, y) \in \mathbb{R}^2 : x \geq 0\}$ on $\mathbb{R}^2$. Also shade in separate sketches the sets $X \cup Y$, $X \cap Y$, $X - Y$, and $Y - X$.

17. Are the following statements true or false?
    (a) $(\mathbb{R} \times \mathbb{Z}) \cap (\mathbb{Z} \times \mathbb{R}) = \mathbb{Z} \times \mathbb{Z}$
    (b) $(\mathbb{R} \times \mathbb{Z}) \cup (\mathbb{Z} \times \mathbb{R}) = \mathbb{R} \times \mathbb{R}$

18. Let $A = \{4, 3, 6, 7, 1, 9\}$ and $B = \{5, 6, 8, 4\}$ have universal set $U = \{0, 1, 2, \ldots, 10\}$. Find:
   (a) $\overline{A}$
   (b) $\overline{B}$
   (c) $A \cap \overline{A}$
   (d) $A \cup \overline{A}$
   (e) $A - \overline{A}$
   (f) $A - \overline{B}$
   (g) $\overline{A} - \overline{B}$
   (h) $\overline{A} \cap B$
   (g) $\overline{\overline{A} \cap B}$

19. Sketch the set $X = [1, 3] \times [1, 2]$ on the plane $\mathbb{R}^2$. Also shade in separate sketches the sets $\overline{X}$ and $\overline{X} \cap ([0, 2] \times [0, 3])$.

20. Sketch the set $X = \{(x, y) \in \mathbb{R}^2 : 1 \le x^2 + y^2 \le 4\}$ on the plane $\mathbb{R}^2$. Also shade in a separate sketch the set $\overline{X}$.

21. Draw a Venn diagram for $\overline{A}$.

22. Draw a Venn diagram for $(A - B) \cap C$.

23. Draw Venn diagrams for $A \cup (B \cap C)$ and $(A \cup B) \cap (A \cup C)$. Based on your drawings, do you think that $A \cup (B \cap C) = (A \cup B) \cap (A \cup C)$?

24. Suppose that sets A and B are in a universal set U. Draw Venn diagrams for $\overline{A \cap B}$ and $\overline{A} \cup \overline{B}$. Based on your drawings, do you think that $\overline{A \cap B} = \overline{A} \cup \overline{B}$?

25. Draw a Venn diagram for $(A \cap B) - C$.

26. Write the corresponding expression for the following Venn diagram involving sets A, B, and C.

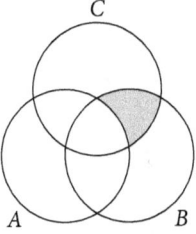

27. Write the corresponding expression for the following Venn diagram involving sets A, B, and C.

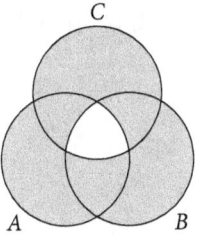

28. Suppose $A_1 = \{a, b, d, e, g, f\}$, $A_2 = \{a, b, c, d\}$, $A_3 = \{b, d, a\}$, and $A_4 = \{a, b, h\}$.

(a) $\bigcup_{i=1}^{4} A_i =$

(b) $\bigcap_{i=1}^{4} A_i =$

29. For each $n \in \mathbb{N}$, let $A_n = \{0, 1, 2, 3, \ldots, n\}$.

(a) $\bigcup_{i \in \mathbb{N}} A_i =$

(b) $\bigcap_{i \in \mathbb{N}} A_i =$

30. (a) $\bigcup_{i \in \mathbb{N}} [i, i+1] =$

(b) $\bigcap_{i \in \mathbb{N}} [i, i+1] =$

31. (a) $\bigcup_{i \in \mathbb{N}} \mathbb{R} \times [i, i+1] =$

(b) $\bigcap_{i \in \mathbb{N}} \mathbb{R} \times [i, i+1] =$

32. (a) $\bigcup_{X \in P(\mathbb{N})} X =$

(b) $\bigcap_{X \in P(\mathbb{N})} X =$

33. Is $\bigcap_{\alpha \in I} A_\alpha \subseteq \bigcup_{\alpha \in I} A_\alpha$ always true for any collection of sets A_α with index set I?

34. If $J \neq \emptyset$ and $J \subseteq I$, does it follow that $\bigcup_{\alpha \in J} A_\alpha \subseteq \bigcup_{\alpha \in I} A_\alpha$? What about $\bigcap_{\alpha \in J} A_\alpha \subseteq \bigcap_{\alpha \in I} A_\alpha$?

2 The Real Number System

Properties of Numbers

You're already familiar with the basic properties of numbers—addition, multiplication, subtraction, division, solutions of equations and inequalities, factoring, and other algebraic manipulations—but this chapter isn't simply a review of old material. Instead, we'll condense this knowledge into a few simple properties, some of which will appear too obvious to mention but on closer inspection will turn out to be surprisingly diverse and subtle.

Of the twelve properties that we'll study in this section, the first nine are concerned with the operations of addition and multiplication. We'll start with addition: this operation is performed on a pair of numbers—the sum $a + b$ exists for any two given numbers a and b (which can be the same number, of course). It might seem reasonable to regard addition as an operation that can be performed on several numbers at the same time, and consider the sum $a_1 + \cdots + a_n$ of n numbers $a_1, \ldots, a_n$ to be a basic concept. It's more convenient, however, to consider addition of only pairs of numbers, and to define other sums in terms of sums of this type. The three numbers a, b, and c, for example, can be summed in two different ways:

- Add b and c, obtaining $b + c$, and then add a to this number to get $a + (b + c)$
- Add a and b, and then add the sum $a + b$ to c to get $(a + b) + c$

Of course, these two compound sums are equal, and this fact is the first of our twelve properties:

(P1)   If a, b, and c are any numbers, then
$$a + (b + c) = (a + b) + c$$

This property renders a separate concept of the sum of three numbers to be superfluous; we simply agree that $a + b + c$ denotes the number $a + (b + c) = (a + b) + c$. Adding four numbers requires similar, though more involved, treatment. The symbol $a + b + c + d$ is defined to mean

(1) $((a + b) + c) + d$ or
(2) $(a + (b + c)) + d$ or
(3) $a + ((b + c) + d)$ or
(4) $a + (b + (c + d))$ or
(5) $(a + b) + (c + d)$

This definition is unambiguous because these five sums are all equal. Fortunately, this fact needn't be listed separately because it follows from the property P1 already given. For example, we know from P1 that

$$(a + b) + c = a + (b + c)$$

and it follows immediately that (1) and (2) are equal. The equality of (2) and (3) is a direct consequence of P1, although this might not be obvious at first glance (let $b + c$ play the role of b in P1, and d play the role of c). The equalities (3) = (4) = (5) are also simple to prove.

It's probably obvious that P1 can be used to prove the equality of the fourteen possible ways to sum five numbers, but it might be less clear how to reasonably arrange this proof without actually listing these fourteen sums. Such a procedure quickly becomes impractical when considering collections of six, seven, or more numbers, and so would be inadequate to prove the equality of all possible sums of an arbitrary finite collection of numbers $a_1, \ldots, a_n$. From now on, we'll disregard the arrangement of parentheses when we write sums $a_1 + \cdots + a_n$. (The proof that the placement of parentheses in a sum is irrelevant involves mathematical induction and isn't included in this book.)

The number 0 has two properties that play crucial roles:

(P2)   If a is any number, then
$$a + 0 = 0 + a = a$$

(P3)   For every number a, there is a number $-a$ such that
$$a + (-a) = (-a) + a = 0$$

Property P2 represents a distinguishing characteristic of the number 0, and (perhaps surprisingly) we can already prove it. If a number x satisfies
$$a + x = a$$
for any one number a, then $x = 0$ (and consequently this equation also holds for all numbers a). The proof of this assertion involves simply subtracting a from both sides of the equation—in other words, adding $-a$ to both sides. As the following proof shows, all three properties P1–P3 must be used to justify this operation.

$$\begin{aligned}
&\text{If} & &a + x = a \\
&\text{then} & &(-a) + (a + x) = (-a) + a = 0 \\
&\text{hence} & &((-a) + a) + x = 0 \\
&\text{hence} & &0 + x = 0 \\
&\text{hence} & &x = 0
\end{aligned}$$

As just hinted, it's convenient to regard subtraction as an operation derived from addition: we consider $a - b$ to be an abbreviation for $a + (-b)$. It's then possible to find the solution of certain simple equations by using a series of steps (each justified by P1, P2, or P3) similar to the ones just given for the equation $a + x = a$. For example:

$$\begin{aligned}
&\text{If} & &x + 3 = 5 \\
&\text{then} & &(x + 3) + (-3) = 5 + (-3) \\
&\text{hence} & &x + (3 + (-3)) = 5 - 3 = 2 \\
&\text{hence} & &x + 0 = 2 \\
&\text{hence} & &x = 2
\end{aligned}$$

Naturally, such elaborate solutions will hold your interest only until you become convinced that they can always be provided. In practice, it's usually a waste of time to solve an equation by explicitly invoking properties P1, P2, and P3 (or any of the upcoming properties).

One other property of addition remains. When we considered the sums of three numbers a, b, and c, only two sums were mentioned: $(a + b) + c$ and $a + (b + c)$. Actually, several other arrangements are possible if a, b, and c are reordered. That these sums are all equal depends on

(P4)   If a and b are any numbers, then
$$a + b = b + a$$

Property P4 emphasizes that although the operation of addition of pairs of numbers might conceivably depend on the order of the two numbers, in fact it does not. Keep in mind that not all operations are so well behaved. Subtraction, for example, lacks this property: usually $a - b \neq b - a$. Elementary algebra shows that $a - b = b - a$ only when $a = b$. Nevertheless, it's impossible to derive this fact from properties P1–P4. We'll be able to justify all the steps needed to show just when $a - b$ equals $b - a$ after we introduce a few more properties and, strangely enough, the crucial property involves multiplication.

The basic properties of multiplication are so similar to those of addition that both their meaning and consequences should be clear. As in elementary algebra, the product of a and b is denoted by $a \cdot b$, or simply ab.

(P5)   If a, b, and c are any numbers, then
$$a \cdot (b \cdot c) = (a \cdot b) \cdot c$$

(P6)   If a is any number, then
$$a \cdot 1 = 1 \cdot a = a$$

Moreover, $1 \neq 0$

(P7)   For every number $a \neq 0$, there is a number a^{-1} such that
$$a \cdot a^{-1} = a^{-1} \cdot a = 1$$

(P8) If a and b are any numbers, then

$$a \cdot b = b \cdot a$$

The assertion that $1 \neq 0$ in property P6 probably strikes you as an odd fact to state, but it's necessary because there's no way to prove it on the basis of the other given properties—these properties would all hold true if there were only one number, namely, 0.

The condition $a \neq 0$ in property P7 is necessary. Because $0 \cdot b = 0$ for all numbers b, there is no number 0^{-1} satisfying $0 \cdot 0^{-1} = 1$. This restriction has an important consequence for division. Just as subtraction was defined in terms of addition, so division is defined in terms of multiplication: the symbol a/b means $a \cdot b^{-1}$. Because 0^{-1} is meaningless, $a/0$ is also meaningless—division by 0 is *always* undefined.

Property P7 has two important consequences. If $a \cdot b = a \cdot c$, then it doesn't necessarily follow that $b = c$; for if $a = 0$, then both $a \cdot b$ and $a \cdot c$ are 0, no matter what b and c are. However, if $a \neq 0$, then $b = c$; this fact can be deduced from P7 as follows:

If $\quad a \cdot b = a \cdot c$ and $a \neq 0$
then $\quad a^{-1} \cdot (a \cdot b) = a^{-1} \cdot (a \cdot c)$
hence $\quad (a^{-1} \cdot a) \cdot b = (a^{-1} \cdot a) \cdot c$
hence $\quad 1 \cdot b = 1 \cdot c$
hence $\quad b = c$

It's also a consequence of P7 that if $a \cdot b = 0$, then either $a = 0$ or $b = 0$. In fact,

If $\quad a \cdot b = 0$ and $a \neq 0$
then $\quad a^{-1} \cdot (a \cdot b) = 0$
hence $\quad (a^{-1} \cdot a) \cdot b = 0$
hence $\quad 1 \cdot b = 0$
hence $\quad b = 0$

It may happen that $a = 0$ and $b = 0$ are both true. This possibility isn't excluded when we say "either $a = 0$ or $b = 0$"; in mathematics, the word "or" is always used in the sense of "one or the other, or both".

This latter consequence of P7 is used constantly in the solution of equations. Suppose, for example, that a number x is known to satisfy

$$(x - 1)(x - 2) = 0$$

It follows that either $x - 1 = 0$ or $x - 2 = 0$; hence $x = 1$ or $x = 2$.

On the basis of the eight properties given so far it's possible to prove very little. The next property, which combines the operations of addition and multiplication, alters this situation dramatically.

(P9) If a, b, and c are any numbers, then

$$a \cdot (b + c) = a \cdot b + a \cdot c$$

(Note that the equation $(b + c) \cdot a = b \cdot a + c \cdot a$ is also true, by P8.)

As an example of the usefulness of P9, we'll show just when $a - b = b - a$:

If	$a - b = b - a$
then	$(a - b) + b = (b - a) + b = b + (b - a)$
hence	$a = b + b - a$
hence	$a + a = (b + b - a) + a = b + b$
consequently	$a \cdot (1 + 1) = b \cdot (1 + 1)$
and therefore	$a = b$

Another use of P9 is the justification of $a \cdot 0 = 0$, which we asserted earlier. This fact wasn't given as one of the basic properties, even though no proof was offered when it was first mentioned. With P1–P8 alone a proof wasn't possible, because the number 0 appears in only P2 and P3, which concern addition, while $a \cdot 0 = 0$ involves multiplication. With P9 the proof is simple, though perhaps not obvious. We have

$$a \cdot 0 + a \cdot 0 = a \cdot (0 + 0)$$
$$= a \cdot 0$$

As we've already noted, this result immediately implies (by adding $-(a \cdot 0)$ to both sides) that $a \cdot 0 = 0$.

P9 can also help explain the somewhat mysterious rule that the product of two negative numbers is positive. First, we'll establish the more easily acceptable assertion that $(-a) \cdot b = -(a \cdot b)$. To prove this, first note that

$$(-a) \cdot b + a \cdot b = [(-a) + a] \cdot b$$
$$= 0 \cdot b$$
$$= 0$$

It follows immediately (by adding $-(a \cdot b)$ to both sides) that $(-a) \cdot b = -(a \cdot b)$. Now note that

$$(-a) \cdot (-b) + [-(a \cdot b)] = (-a) \cdot (-b) + (-a) \cdot b$$
$$= (-a) \cdot [(-b) + b]$$
$$= (-a) \cdot 0$$
$$= 0$$

Consequently, adding $(a \cdot b)$ to both sides, we get

$$(-a) \cdot (-b) = a \cdot b$$

The fact that the product of two negative numbers is positive is thus a consequence of P1–P9. In other words, crucially, if we want P1 to P9 to be true, then the rule for the product of two negative numbers is forced on us.

The various consequences of P9 examined so far, although interesting and important, don't really indicate the significance of P9—after all, we could have listed each of these properties separately. Actually, P9 is the justification for almost all algebraic manipulations. For example, although we've shown how to solve the equation

$$(x - 1)(x - 2) = 0$$

we're more likely to encounter the equation in the form

$$x^2 - 3x + 2 = 0$$

The factorization $x^2 - 3x + 2 = (x - 1)(x - 2)$ is really a triple use of P9:

$$\begin{aligned}
(x - 1) \cdot (x - 2) &= x \cdot (x - 2) + (-1) \cdot (x - 2) \\
&= x \cdot x + x \cdot (-2) + (-1) \cdot x + (-1) \cdot (-2) \\
&= x^2 + x[(-2) + (-1)] + 2 \\
&= x^2 - 3x + 2
\end{aligned}$$

A final illustration of the importance of P9 is the fact that this property is actually used every time one multiplies arabic numerals. For example, the calculation

$$\begin{array}{r} 13 \\ \times 24 \\ \hline 52 \\ 26 \\ \hline 312 \end{array}$$

is actually a concise arrangement for the following equations:

$$\begin{aligned}
13 \cdot 24 &= 13 \cdot (2 \cdot 10 + 4) \\
&= 13 \cdot 2 \cdot 10 + 13 \cdot 4 \\
&= 26 \cdot 10 + 52
\end{aligned}$$

Note that moving 26 to the left in the above calculation is the same as writing $26 \cdot 10$. The multiplication $13 \cdot 4 = 52$ uses P9 also:

$$\begin{aligned}
13 \cdot 4 &= (1 \cdot 10 + 3) \cdot 4 \\
&= 1 \cdot 10 \cdot 4 + 3 \cdot 4 \\
&= 4 \cdot 10 + 12 \\
&= 4 \cdot 10 + 1 \cdot 10 + 2 \\
&= (4 + 1) \cdot 10 + 2 \\
&= 5 \cdot 10 + 2 \\
&= 52
\end{aligned}$$

The properties P1–P9 have descriptive names which aren't essential to remember, but which are convenient for reference. Here are properties P1–P9 listed with the names by which they're commonly designated.

P1—Associative law for addition

$$a + (b + c) = (a + b) + c$$

P2—Existence of an additive identity

$$a + 0 = 0 + a = a$$

P3—Existence of additive inverses

$$a + (-a) = (-a) + a = 0$$

P4—Commutative law for addition

$$a + b = b + a$$

P5—Associative law for multiplication

$$a \cdot (b \cdot c) = (a \cdot b) \cdot c$$

P6—Existence of a multiplicative identity

$$a \cdot 1 = 1 \cdot a = a; \quad 1 \neq 0$$

P7—Existence of multiplicative inverses

$$a \cdot a^{-1} = a^{-1} \cdot a = 1, \quad \text{for } a \neq 0$$

P8—Commutative law for multiplication

$$a \cdot b = b \cdot a$$

P9—Distributive law

$$a \cdot (b + c) = a \cdot b + a \cdot c$$

The three basic properties of numbers which remain to be listed are concerned with inequalities. Although inequalities occur rarely in elementary mathematics, they play a prominent role in calculus. The two notions of inequality, $a < b$ (a is less than b) and $a > b$ (a is greater than b), are intimately related: $a < b$ means the same as $b > a$ (thus $1 < 3$ and $3 > 1$ are merely two ways of writing the same assertion). The numbers a satisfying $a > 0$ are called **positive**, while those numbers a satisfying $a < 0$ are called **negative**. While positivity can thus be defined in terms of $<$, it's possible to reverse the procedure: $a < b$ can be defined to mean that $b - a$ is positive. In fact, it's convenient to consider the collection of all positive numbers, denoted by P, as the basic concept, and state all properties in terms of P:

(P10) (Trichotomy law) For every number a, one and only one of the following holds:
  (i) $a = 0$
  (ii) a is in the collection P
  (iii) $-a$ is in the collection P

(P11) (Closure under addition) If a and b are in P, then $a + b$ is in P.

(P12) (Closure under multiplication) If a and b are in P, then $a \cdot b$ is in P.

These three properties are complemented with the following definitions:

$$a > b \quad \text{if} \quad a - b \text{ is in } P$$
$$a < b \quad \text{if} \quad b > a$$
$$a \geq b \quad \text{if} \quad a > b \text{ or } a = b$$
$$a \leq b \quad \text{if} \quad a < b \text{ or } a = b$$

Note in particular that $a > 0$ if and only if a is in P.

All the familiar facts about inequalities, however elementary they might seem to be, are consequences of P10–P12. For example, if a and b are any two numbers, then exactly one of the following holds:

  (i) $a - b = 0$
  (ii) $a - b$ is in the collection P
  (iii) $-(a - b) = b - a$ is in the collection P

Using the definitions just made, it follows that exactly one of the following holds:

(i) $a = b$
(ii) $a > b$
(iii) $b > a$

A more interesting fact results from the following manipulations. If $a < b$, so that $b - a$ is in P, then surely $(b + c) - (a + c)$ is in P; thus, if $a < b$, then $a + c < b + c$. Similarly, suppose that $a < b$ and $b < c$,

then $b - a$ is in P
and $c - b$ is in P
so $c - a = (c - b) + (b - a)$ is in P

This result shows that if $a < b$ and $b < c$, then $a < c$. (The two inequalities $a < b$ and $b < c$ are usually written in the abbreviated form $a < b < c$, which has the built-in third inequality $a < c$.)

The following assertion is less obvious: if $a < 0$ and $b < 0$, then $ab > 0$. The difficulty in the proof is the unraveling of definitions. The symbol $a < 0$ means, by definition, $0 > a$, which means $0 - a = -a$ is in P. Similarly $-b$ is in P, and consequently, by P12, $(-a)(-b) = ab$ is in P. Thus $ab > 0$.

The fact that $ab > 0$ if $a > 0$, $b > 0$ and also if $a < 0$, $b < 0$ has one special consequence: $a^2 > 0$ if $a \neq 0$. Thus squares of nonzero numbers are always positive, and so we've just proved a result which might have seemed sufficiently elementary to be included in our list of properties: $1 > 0$ (because $1 = 1^2$).

We'll note one slightly perplexing feature of the symbols $\geq$ and $\leq$. The statements

$$1 + 1 \leq 3$$
$$1 + 1 \leq 2$$

are both true, even though we know that $\leq$ could be replaced by $<$ in the first statement, and by $=$ in the second. This sort of thing is bound to occur when $\leq$ is used with specific numbers; the usefulness of the symbol is revealed by a statement like Theorem 2.1 in the next section, in which equality holds for some values of a and b, while inequality holds for other values.

We'll conclude this section with a subtle yet important point that was glossed over earlier. After stating property P9, we proved that $a - b = b - a$ implies $a = b$. That proof established that

$$a \cdot (1 + 1) = b \cdot (1 + 1)$$

from which we concluded that $a = b$. This result is obtained from the equation $a \cdot (1 + 1) = b \cdot (1 + 1)$ by dividing both sides by $1 + 1$. Division by 0 is undefined, so the validity of the argument depends on knowing that $1 + 1 \neq 0$. This fact can't be proved from properties P1–P9 alone (see Problem 18); P10, P11, and P12 are needed. The proof is trivially simple, however: we've already seen that $1 > 0$; it follows that $1 + 1 > 0$, and in particular $1 + 1 \neq 0$.

The preceding proof may have confirmed your suspicion that it's silly to bother proving such obvious facts, but an honest assessment of the situation will help justify serious consideration of such details. In this section we've assumed that numbers are familiar objects, and that P1–P12 are merely statements of obvious, well-known properties of numbers. It would be difficult, however, to justify this assumption. Although you learned how to "work with" numbers in school, just what numbers *are* remains shrouded in darkness. This book aims to elucidate the concept of numbers, and by the end you'll be well-acquainted with them, though not *thoroughly* acquainted with them—that takes a lifetime. This understanding hinges on your diligent attempt to answer the problems, many of which are designed to extend your knowledge rather than merely reinforce it. In fact, by now you should be convinced that P1–P12 are indeed basic properties which we should assume to deduce other familiar properties of numbers (note that P1–P12 don't account for *all* properties of numbers, and later sections point out some deficiencies of P1–P12).

Absolute Value

The fact that $-a > 0$ if $a < 0$ is the basis of a concept that plays a crucial role in mathematics. For any number a, we define the **absolute value** $|a|$ of a as follows:

$$|a| = \begin{cases} a, & a \geq 0 \\ -a, & a < 0 \end{cases}$$

Note that $|a|$ is always positive, except when $a = 0$. For example, we have $|-3| = 3$, $|7| = 7$, $|1 + \sqrt{2} - \sqrt{3}| = 1 + \sqrt{2} - \sqrt{3}$, and $|1 + \sqrt{2} - \sqrt{10}| = \sqrt{10} - \sqrt{2} - 1$. In general, the most straightforward approach to any problem involving absolute values is to treat several cases separately, since absolute values are defined by cases to begin with. This approach can be used to prove the following important fact about absolute values.

Theorem 2.1 For all numbers a and b, we have

$$|a + b| \leq |a| + |b|$$

Proof We'll consider four cases:

(1) $a \geq 0, \ b \geq 0$
(2) $a \geq 0, \ b \leq 0$
(3) $a \leq 0, \ b \geq 0$
(4) $a \leq 0, \ b \leq 0$

In case (1) we also have $a + b \geq 0$, and the theorem obviously holds true here; in fact,

$$|a + b| = a + b = |a| + |b|$$

so that in this case equality holds.

In case (4) we have $a + b \leq 0$, and the theorem holds true here as well:

$$|a + b| = -(a + b) = -a + (-b) = |a| + |b|$$

In case (2), when $a \geq 0$ and $b \leq 0$, we must prove that

$$|a + b| \leq a - b$$

This case can therefore be divided into two subcases: $a + b \geq 0$ and ≤ 0.

If $a + b \geq 0$, then we must prove that

$$a + b \leq a - b$$

that is,

$$b \leq -b$$

which is certainly true because $b \leq 0$ and hence $-b \geq 0$. In the other subcase, if $a + b \leq 0$, then we must prove that

$$-a - b \leq a - b$$

that is,

$$-a \leq a$$

which is certainly true because $a \geq 0$ and hence $-a \leq 0$.

Finally, to show that case (3) is true, simply apply case (2) with a and b interchanged. ∎

Although this method of treating absolute values (considering various cases separately) is sometimes the only approach available, simpler methods can often be used. In fact, it's possible to give a much shorter proof of Theorem 2.1. This proof is motivated by the observation that

$$|a| = \sqrt{a^2}$$

(Here, and throughout this book, $\sqrt{x}$ denotes the *positive* square root of x; this symbol is defined only if $x \geq 0$.)

We can now observe that

$$\begin{aligned}(|a + b|)^2 &= (a + b)^2 \\ &= a^2 + 2ab + b^2 \\ &\leq a^2 + 2|a| \cdot |b| + b^2 \\ &= |a|^2 + 2|a| \cdot |b| + |b|^2 \\ &= (|a| + |b|)^2\end{aligned}$$

From this result we can conclude that $|a + b| \leq |a| + |b|$ because $x^2 < y^2$ implies $x < y$, provided that x and y are both nonnegative (see Problem 5).

One final observation about the theorem we've just proved: a close examination of either proof given shows that

$$|a + b| = |a| + |b|$$

if a and b have the same sign (that is, are both positive or both negative), or if one of the two is 0, whereas

$$|a + b| < |a| + |b|$$

if a and b are of opposite signs.

Natural Numbers and Induction

To this point we've used the word "number" loosely, despite our concern with the basic properties of numbers. It's now necessary to distinguish various types of numbers.

The simplest numbers are the counting numbers

$$1, 2, 3, \ldots$$

The fundamental significance of this collection of numbers is emphasized by its symbol $\mathbb{N}$ (for **natural numbers**). A brief glance at properties P1–P12 will show that our basic properties of "numbers" don't apply to $\mathbb{N}$—for example, P2 and P3 don't make sense for $\mathbb{N}$ because $\mathbb{N}$ lacks the number 0. From this point of view the system $\mathbb{N}$ has many deficiencies. Nevertheless, $\mathbb{N}$ is sufficiently important to deserve comments before we consider larger collections of numbers.

The most basic property of $\mathbb{N}$ is the principle of mathematical induction. Suppose that $P(x)$ means that the property P holds for the number x. Then the principle of **mathematical induction** states that $P(x)$ is true for all natural numbers x provided that

(1) $P(1)$ is true.

(2) Whenever $P(k)$ is true, $P(k + 1)$ is true.

Note that condition (2) merely asserts the truth of $P(k + 1)$ under the assumption that $P(k)$ is true; this condition suffices to ensure the truth of $P(x)$ for all x, if condition (1) also holds. In fact, if $P(1)$ is true, then it follows that $P(2)$ is true (by using (2) in the special case $k = 1$).

Now, because P(2) is true it follows that P(3) is true (by using (2) in the special case $k = 2$). It's clear that each number will eventually be reached by a series of steps of this sort, so that P(k) is true for all numbers k.

A common illustration of the reasoning behind mathematical induction envisions an infinite line of people,

person number 1, person number 2, person number 3,...

If each person has been instructed to tell any secret he hears to the person behind him (the one with the next largest number) and a secret is told to person number 1, then clearly every person will eventually learn the secret. If $P(x)$ is the assertion that person number x will learn the secret, then the instructions given (to tell all secrets learned to the next person) assures that condition (2) is true, and telling the secret to person number 1 makes (1) true. The following example is a more-serious use of mathematical induction. A useful and striking formula expresses the sum of the first n numbers in a simple way:

$$1 + \cdots + n = \frac{n(n+1)}{2}$$

To prove this formula, note first that it's clearly true for $n = 1$. Now *assume* that for some natural number k we have

$$1 + \cdots + k = \frac{k(k+1)}{2}$$

Then

$$1 + \cdots + k + (k+1) = \frac{k(k+1)}{2} + k + 1$$
$$= \frac{k(k+1) + 2k + 2}{2}$$
$$= \frac{k^2 + 3k + 2}{2}$$
$$= \frac{(k+1)(k+2)}{2}$$

so the formula is also true for $k + 1$. By the principle of induction this proves the formula for all natural numbers n. This particular example illustrates a phenomenon that occurs frequently, especially in connection with formulas like the one just proved. Problem 24 shows how a formula of this type can be derived.

The principle of mathematical induction can be formulated in an equivalent way without speaking of "properties" of a number, a term that's sufficiently vague to be avoided in mathematical discussions. A more precise formulation states that if A is any collection (or set—a synonymous mathematical term) of natural numbers and

(1) 1 is in A,

(2) $k+1$ is in A whenever k is in A,

then A is the set of all natural numbers. It should be clear that this formulation adequately replaces the less formal one given previously—we just consider the set A of natural numbers x which satisfy $P(x)$. For example, suppose that A is the set of natural numbers n for which it's true that

$$1+\cdots+n = \frac{n(n+1)}{2}$$

Our previous proof of this formula showed that A contains 1, and that $k+1$ is in A, if k is. It follows that A is the set of all natural numbers; that is, that the formula holds for all natural numbers n.

There's yet another rigorous formulation of the principle of mathematical induction, which looks quite different. If A is any collection of natural numbers, then it's tempting to say that A must have a smallest member. Actually, this statement can fail to be true in a subtle way. An important set of natural numbers is the collection A that contains no natural numbers at all, the **empty set** or **null set**, denoted by $\emptyset$. The null set $\emptyset$ is a collection of natural numbers that has no smallest member—in fact, it has no members at all. This set is the only possible exception, however; if A is a non-null set of natural numbers, then A has a least member. This "intuitively obvious" statement, known as the **well-ordering principle**, can be proved from the principle of induction as follows. Suppose that the set A has no least member. Let B be the set of natural numbers n such that $1,\ldots,n$ are all *not* in A. Clearly 1 is in B (because if 1 were in A, then A would have 1 as its smallest member).

Moreover, if $1, \ldots, k$ are not in A, surely $k + 1$ is not in A (otherwise $k + 1$ would be the smallest member of A), so $1, \ldots, k + 1$ are all not in A. This reasoning shows that if k is in B, then $k + 1$ is in B. It follows that every number n is in B; that is, the numbers $1, \ldots, n$ are not in A for any natural number n. Thus $A = \emptyset$, which completes the proof.

It's also possible to prove the principle of induction from the well-ordering principle. Suppose that A contains 1, and that A contains $n + 1$ if it contains n. If A does not contain all natural numbers, then the set B of natural numbers *not* in A is not $\emptyset$. So B has a smallest member n_0. Now $n_0 \neq 1$, because A contains 1, so we can write $n_0 = (n_0 - 1) + 1$, where $n_0 - 1$ is a natural number. Now $n_0 - 1$ is *not* in B, so $n_0 - 1$ is in A. By hypothesis, n_0 must be in A, so n_0 is not in B, a contradiction. (By the way, the assertion that a natural number $n \neq 1$ can be written $n = m + 1$ for some other natural number m, can itself be proved by induction.)

Yet another form of induction must be mentioned. It sometimes happens that to prove $P(k + 1)$ we must assume not only $P(k)$, but also $P(l)$ for all natural numbers $l \leq k$. In this case we rely on the **principle of complete induction**: if A is a set of natural numbers and

(1) 1 is in A,

(2) $k + 1$ is in A if $1, \ldots, k$ are in A,

then A is the set of all natural numbers.

Although the principle of complete induction may appear stronger than the ordinary principle of induction, it's actually a consequence of that principle, and one can prove the principle of complete induction from the ordinary principle of induction. If A contains 1 and A contains $n + 1$ whenever it contains $1, \ldots, n$, then consider the set B of all k such that $1, \ldots, k$ are all in A. Clearly 1 is in B. If k is in B, then $1, \ldots, k$ are all in A, so $k + 1$ is in A, so $1, \ldots, k + 1$ are in A, so $k + 1$ is in B. By (ordinary) induction, $B = \mathbb{N}$, so also $A = \mathbb{N}$.

Closely related to proofs by induction are **recursive definitions**. For example, the number $n!$ (read "n factorial") is defined as the product of all the natural numbers less than or equal to n:

$$n! = 1 \cdot 2 \cdots (n - 1) \cdot n$$

This definition can be expressed more precisely as follows:

(1) $1! = 1$

(2) $n! = n \cdot (n-1)!$

This form of the definition exhibits the relationship between $n!$ and $(n-1)!$ in an explicit way that's ideally suited for proofs by induction. Problem 31 reviews a definition already familiar to you, which can be expressed more succinctly as a recursive definition; as this problem shows, the recursive definition is necessary for a rigorous proof of some of the basic properties of the definition.

One final note regarding the null set: although it may not strike you as a collection in the ordinary sense of the word, the null set arises naturally in many contexts. We frequently consider the set A, consisting of all x satisfying some property P. Often we have no guarantee that P is satisfied by any number, so that A might be $\emptyset$—in fact, often one proves that P is always false by showing that $A = \emptyset$.

Summation

One definition that may not be familiar involves some convenient and frequently used **sigma notation**. Instead of writing

$$a_1 + a_2 + \cdots + a_n$$

mathematicians usually use the Greek letter Σ (capital sigma, for *sum*) and write

$$\sum_{i=1}^{n} a_i$$

In other words, $\sum_{i=1}^{n} a_i$ denotes the sum of the numbers obtained by letting $i = 1, 2, \ldots, n$. Thus

$$\sum_{i=1}^{n} i = 1 + 2 + \cdots + n = \frac{n(n+1)}{2}$$

Note that the letter i really has nothing to do with the number denoted by $\sum_{i=1}^{n} i$ and can be replaced by any convenient symbol (except n, of course). Some equivalent symbols are:

$$\sum_{j=1}^{n} j = \frac{n(n+1)}{2}$$

$$\sum_{j=1}^{i} j = \frac{i(i+1)}{2}$$

$$\sum_{n=1}^{j} n = \frac{j(j+1)}{2}$$

To define $\sum_{i=1}^{n} a_i$ precisely actually requires a recursive definition:

(1) $\sum_{i=1}^{1} a_i = a_1$

(2) $\sum_{i=1}^{n} a_i = \sum_{i=1}^{n-1} a_i + a_n$

But only mathematical sticklers would strongly insist on such precision. In practice, all sorts of modifications of this symbolism are used, and no one ever considers it necessary to add any words of explanation. For example, the symbol

$$\sum_{\substack{i=1 \\ i \neq 4}}^{n} a_i$$

is an obvious way of writing

$$a_1 + a_2 + a_3 + a_5 + a_6 + \cdots + a_n$$

or more precisely

$$\sum_{i=1}^{3} a_i + \sum_{i=5}^{n} a_i$$

Integers, Rationals, and Reals

The deficiencies of the natural numbers that we discovered earlier can be partially remedied by extending this system to the set of **integers**

$$\ldots, -3, -2, -1, 0, 1, 2, 3, \ldots$$

This set is denoted by $\mathbb{Z}$ (for "Zahl", German for "number"). Of properties P1–P12, only P7 fails for $\mathbb{Z}$.

A still larger system of numbers is obtained by taking quotients m/n of integers (with $n \neq 0$). These numbers are called **rational numbers**, and the set of all rational numbers is denoted by $\mathbb{Q}$ (for "quotients"). In this system of numbers all of P1–P12 are true. It's tempting to conclude that the properties of numbers refer to only one set of numbers, namely, $\mathbb{Q}$.

There is, however, a still larger collection of numbers to which properties P1–P12 apply—the set of all **real numbers**, denoted by $\mathbb{R}$. The real numbers include not only the rational numbers but also the **irrational numbers**, which can be represented by infinite decimals. π and $\sqrt{2}$ are both examples of irrational numbers. The proof that π is irrational is beyond the scope of this book. The irrationality of $\sqrt{2}$, on the other hand, is simple, and was known to the ancient Greeks. (Because the Pythagorean theorem shows that an isosceles right triangle, with legs of length 1, has a hypotenuse of length $\sqrt{2}$, it's not surprising that the Greeks investigated this question.) The proof depends on a few observations about the natural numbers. Every natural number n can be written either in the form $2k$ for some integer k, or else in the form $2k + 1$ for some integer k (see Problem 26 for a proof by induction). Those natural numbers of the form $2k$ are called **even**; those of the form $2k + 1$ are called **odd**. Note that even numbers have even squares, and odd numbers have odd squares:

$$(2k)^2 = 4k^2 = 2 \cdot (2k^2)$$

$$(2k + 1)^2 = 4k^2 + 4k + 1 = 2 \cdot (2k^2 + 2k) + 1$$

In particular it follows that the converse must also hold: if n^2 is even, then n is even; if n^2 is odd, then n is odd. The proof that $\sqrt{2}$ is irrational is now simple. Suppose that $\sqrt{2}$ were rational; that is, suppose there were natural numbers p and q such that

$$\left(\frac{p}{q}\right)^2 = 2$$

We can assume that p and q have no common divisor (because all common divisors could be divided out to begin with). Now we have

$$p^2 = 2q^2$$

This result shows that p^2 is even, and consequently p must be even; that is, $p = 2k$ for some natural number k. Then

$$p^2 = 4k^2 = 2q^2$$

so

$$2k^2 = q^2$$

This result shows that q^2 is even, and consequently that q is even. Thus both p and q are even, contradicting the fact that p and q have no common divisor. This contradiction completes the proof.

It's important to understand precisely what this proof shows. We've demonstrated that there is no rational number x such that $x^2 = 2$. This assertion is often expressed more briefly by saying that $\sqrt{2}$ is irrational. Note, however, that the use of the symbol $\sqrt{2}$ implies the existence of *some* number (necessarily irrational) whose square is 2. We haven't proved that such a number exists and we can assert confidently that, at present, a proof is *impossible* for us. Any proof at this stage would have to be based on properties P1–P12 (the only properties of $\mathbb{R}$ that we've mentioned); because P1–P12 are also true for $\mathbb{Q}$, the exact same argument would show that there's a rational number whose square is 2, and this we know is false. Note that the reverse argument won't work—our proof that there's no rational number whose square is 2 can't be used to show that there is no real number whose square is 2, because our proof used not only P1–P12 but also a special property of $\mathbb{Q}$, the fact that every number in $\mathbb{Q}$ can be written p/q for integers p and q.

This particular deficiency in our list of properties of the real numbers could, of course, be corrected by adding a new property that asserts the existence of square roots of positive numbers. Resorting to such a measure is, however, clumsy and mathematically unsatisfactory; we'd still not know that every number has an nth root if n is odd, and that every positive number has an nth root if n is even. Even if we assumed this, we couldn't prove the existence of a number x satisfying $x^5 + x + 1 = 0$ (even though there does happen to be one), because we don't know how to write the solution of the equation in terms of nth roots (in fact, it's known that the solution can't be written in this form). And, of course, we certainly don't want to falsely assume that all equations have solutions (no real number x satisfies $x^2 + 1 = 0$, for example). In fact, this direction of investigation is fruitless. Hence we arrive at this book's natural stopping point; the property distinguishing $\mathbb{R}$ from $\mathbb{Q}$ comes not from the study of real numbers alone, but the advanced study of the foundations of calculus.

Problems

1. Prove the following:

    (a) If $ax = a$ for some number $a \neq 0$, then $x = 1$.

    (b) $x^2 - y^2 = (x - y)(x + y)$.

    (c) If $x^2 = y^2$, then $x = y$ or $x = -y$.

    (d) $x^3 - y^3 = (x - y)(x^2 + xy + y^2)$.

    (e) $x^n - y^n = (x - y)(x^{n-1} + x^{n-2}y + \cdots + xy^{n-2} + y^{n-1})$.

    (f) $x^3 + y^3 = (x + y)(x^2 - xy + y^2)$. (Hint: There's an easy way to prove this by using part (d), and it will show you how to find a factorization for $x^n + y^n$ whenever n is odd.)

2. What is wrong with the following "proof"? Let $x = y$. Then
$$x^2 = xy$$
$$x^2 - y^2 = xy - y^2$$
$$(x+y)(x-y) = y(x-y)$$
$$x + y = y$$
$$2y = y$$
$$2 = 1$$

3. Prove the following:

    (a) $(ab)^{-1} = a^{-1}b^{-1}$, if $a, b \neq 0$. (Hint: Recall the defining property of $(ab)^{-1}$.)

    (b) $\dfrac{a}{b} = \dfrac{ac}{bc}$, if $b, c \neq 0$.

    (c) $\dfrac{a}{b} + \dfrac{c}{d} = \dfrac{ad + bc}{bd}$, if $b, d \neq 0$.

    (d) $\dfrac{a}{b} \cdot \dfrac{c}{d} = \dfrac{ac}{bd}$, if $b, d \neq 0$.

    (e) $\dfrac{a}{b} \bigg/ \dfrac{c}{d} = \dfrac{ad}{bc}$, if $b, c, d \neq 0$.

    (f) If $b, d \neq 0$, then $\dfrac{a}{b} = \dfrac{c}{d}$ if and only if $ad = bc$. Also determine when $\dfrac{a}{b} = \dfrac{b}{a}$.

4. Find all numbers x for which

    (a) $4 - x < 3 - 2x$.

    (b) $5 - x^2 < 8$.

    (c) $5 - x^2 < -2$.

    (d) $(x - 1)(x - 3) > 0$. (Hint: When is a product of two numbers positive?)

    (e) $x^2 - 2x + 2 > 0$.

    (f) $x^2 + x^2 + 1 > 2$.

(g) $x^2 - x + 10 > 16$.

(h) $x^2 + x + 1 > 0$.

(i) $(x - \pi)(x + 5)(x - 3) > 0$.

(j) $(x - \sqrt[3]{2})(x - \sqrt{2}) > 0$.

(k) $2^x < 8$.

(l) $x + 3^x < 4$.

(m) $\dfrac{1}{x} + \dfrac{1}{1-x} > 0$.

(n) $\dfrac{x-1}{x+1} > 0$.

5. Prove the following:

   (a) If $a < b$ and $c < d$, then $a + c < b + d$.

   (b) If $a < b$, then $-b < -a$.

   (c) If $a < b$ and $c > d$, then $a - c < b - d$.

   (d) If $a < b$ and $c > 0$, then $ac < bc$.

   (e) If $a < b$ and $c < 0$, then $ac > bc$.

   (f) If $a > 1$, then $a^2 > a$.

   (g) If $0 < a < 1$, then $a^2 < a$.

   (h) If $0 \leq a < b$ and $0 \leq c < d$, then $ac < bd$.

   (i) If $0 \leq a < b$, then $a^2 < b^2$. (Hint: Use part (h).)

   (j) If $a, b \geq 0$ and $a^2 < b^2$, then $a < b$. (Hint: Use part (i) backward.)

6. (a) Prove that if $0 \leq x < y$, then $x^n < y^n$, $n = 1, 2, 3, \ldots$.

   (b) Prove that if $x < y$ and n is odd, then $x^n < y^n$.

   (c) Prove that if $x^n = y^n$ and n is odd, then $x = y$.

   (d) Prove that if $x^n = y^n$ and n is even, then $x = y$ or $x = -y$.

7. Express each of the following with at least one fewer pair of absolute value signs.

   (a) $\left|\sqrt{2}+\sqrt{3}-\sqrt{5}+\sqrt{7}\right|$.

   (b) $|(|a+b|-|a|-|b|)|$.

   (c) $|(|a+b|+|c|-|a+b+c|)|$.

   (d) $|x^2-2xy+y^2|$.

   (e) $\left|\left(\left|\sqrt{2}+\sqrt{3}\right|-\left|\sqrt{5}-\sqrt{7}\right|\right)\right|$.

8. Express each of the following without absolute value signs, treating various cases separately when necessary.

   (a) $|a+b|-|b|$.

   (b) $|(|x|-1)|$.

   (c) $|x|-|x^2|$.

   (d) $a-|(a-|a|)|$.

9. Find all numbers x for which

   (a) $|x-3|=8$.

   (b) $|x-3|<8$.

   (c) $|x+4|<2$.

   (d) $|x-1|+|x-2|>1$.

   (e) $|x-1|+|x+1|<2$.

   (f) $|x-1|+|x+1|<1$.

   (g) $|x-1|\cdot|x+1|=0$.

   (h) $|x-1|\cdot|x+2|=3$.

10. Prove the following:

    (a) $|xy|=|x|\cdot|y|$.

    (b) $\left|\dfrac{1}{x}\right|=\dfrac{1}{|x|}$, if $x\neq 0$. (Hint: Recall what $|x|^{-1}$ is.)

Essential Advanced Precalculus

(c) $\left|\dfrac{x}{y}\right| = \dfrac{|x|}{|y|}$, if $y \neq 0$.

(d) $|x - y| \leq |x| + |y|$. (Hint: A very short proof is possible.)

(e) $|x| - |y| \leq |x - y|$. (Hint: A very short proof is possible if you write things in the correct way.)

(f) $|(|x| - |y|)| \leq |x - y|$. (Hint: The proof follows immediately from (e).)

(g) $|x + y + z| \leq |x| + |y| + |z|$. Indicate when equality holds, and prove your statement.

11. Prove that $|a| = |-a|$. (Hint: Don't become confused by too many cases. First prove the statement for $a \geq 0$. Why is it then obvious for $a \leq 0$?)

12. Prove that $-b \leq a \leq b$ if and only if $|a| \leq b$. In particular, it follows that $-|a| \leq a \leq |a|$.

13. Use the result of the preceding problem to give a new proof that $|a + b| \leq |a| + |b|$.

14. Find the smallest possible value of $2x^2 - 3x + 4$. (Hint: Complete the square; that is, write $2x^2 - 3x + 4 = 2(x - \tfrac{3}{4})^2 + ?$.)

15. Find the smallest possible value of $x^2 - 3x + 2y^2 + 4y + 2$.

16. Find the smallest possible value of $x^2 + 4xy + 5y^2 - 4x - 6y + 7$.

17. (a) Suppose that $b^2 - 4c \geq 0$. Show that the numbers
$$\dfrac{-b + \sqrt{b^2 - 4c}}{2}, \quad \dfrac{-b - \sqrt{b^2 - 4c}}{2}$$
both satisfy the equation $x^2 + bx + c = 0$.

(b) Suppose that $b^2 - 4c < 0$. Show that there are no numbers x satisfying $x^2 + bx + c = 0$. In fact, $x^2 + bx + c > 0$ for all x. (Hint: Complete the square.)

(c) Apply the result of part (b) to give another proof that if x and y aren't both 0, then $x^2 + xy + y^2 > 0$.

(d) For which numbers α is it true that $x^2 + \alpha xy + y^2 > 0$ whenever x and y aren't both 0?

18. Suppose that we interpret "number" to mean either 0 or 1, and $+$ and $\cdot$ to be the operations defined by the following two tables. Check that properties P1–P9 all hold, even though $1 + 1 = 0$.

19. Prove the following formula by induction:
$$1^2 + \cdots + n^2 = \frac{n(n+1)(2n+1)}{6}$$

20. Prove the following formula by induction:
$$1^3 + \cdots + n^3 = (1 + \cdots + n)^2$$

21. Find a formula for
$$\sum_{i=1}^{n}(2i-1) = 1 + 3 + 5 + \cdots + (2n-1)$$
(Hint: What does this expression have to do with $1 + 2 + 3 + \cdots + 2n$?)

22. Find a formula for
$$\sum_{i=1}^{n}(2i-1)^2 = 1^2 + 3^2 + 5^2 + \cdots + (2n-1)^2$$
(Hint: What does this expression have to do with $1^2 + 2^2 + 3^2 + \cdots + (2n)^2$?)

23. If $0 \le k \le n$, the **binomial coefficient** $\binom{n}{k}$ is defined by
$$\binom{n}{k} = \frac{n!}{k!(n-k)!} = \frac{n(n-1)\cdots(n-k+1)}{k!}, \quad k \ne 0, n$$
If we define $0! = 1$, then a special case of this formula is
$$\binom{n}{0} = \binom{n}{n} = 1$$
Prove that
$$\binom{n+1}{k} = \binom{n}{k-1} + \binom{n}{k}$$

The proof doesn't require induction.

24. Prove by induction on n that
$$1+r+r^2+\cdots+r^n = \frac{1-r^{n+1}}{1-r}$$
if $r \neq 1$ (if $r = 1$, evaluating the sum is simple).

25. Derive the result of the preceding problem by setting
$$S = 1 + r + \cdots + r^n,$$
multiplying this equation by r, and then solving the two equations for S.

26. Prove that every natural number is either even or odd.

27. If a is rational and b is irrational, is $a + b$ necessarily irrational? What if a and b are both irrational?

28. If a is rational and b is irrational, is ab necessarily irrational? (Be careful.)

29. Is there a number a such that a^2 is irrational, but a^4 is rational?

30. Are there two irrational numbers whose sum and product are both rational?

31. The following is a recursive definition of a^n:
$$a^1 = a$$
$$a^{n+1} = a^n \cdot a$$

Prove by induction that
$$a^{n+m} = a^n \cdot a^m$$
$$(a^n)^m = a^{nm}$$

Use either induction on n or induction on m, not both at the same time.

3 Functions

The Concept of a Function

Functions are the central objects of investigation in almost every branch of modern mathematics. Consequently, the concept of a function is one of vast generality, so for now we'll narrow our attention to a particular class of functions, yet one that exhibits enough variety to discuss functions at length. Later we'll define functions formally, but we'll begin with a provisional definition to illustrate the intuitive notion of functions.

Provisional Definition A **function** is a rule that assigns, to each of certain real numbers, some other real number.

This definition, which obviously needs to be clarified, is illustrated by the following examples.

Example 1
 The rule that assigns to each number the square of that number.

Example 2
 The rule that assigns to each number y the number $\dfrac{y^3+3y+5}{y^2+1}$.

Example 3
 The rule that assigns to each number $c \neq 1, -1$ the number $\dfrac{c^3+3c+5}{c^2-1}$.

Example 4
 The rule that assigns to each number x satisfying $-17 \leq x \leq \pi/3$ the number x^2.

Example 5

The rule that assigns to each number a the number 0 if a is irrational, and the number 1 if a is rational.

Example 6

The rule that assigns

to 2 the number 5,

to 17 the number $36/\pi$,

to $\pi^2/17$ the number 28,

to $36/\pi$ the number 28,

and to any $y \neq 2, 17, \pi^2/17$, or $36/\pi$, the number 16 if y is of the form $a + b\sqrt{2}$ for a, b in $\mathbb{Q}$. [$\mathbb{Q}$ (for "quotients") denotes the set of all rational numbers m/n, where m and n are integers (with $n \neq 0$).]

Example 7

The rule that assigns to each number t the number $t^3 + x$. (This rule clearly depends on the value of x, so it's really describing infinitely many different functions, one for each number x.)

Example 8

The rule that assigns to each number z the number of 7's in the decimal expansion of z, if this number is finite, and $-\pi$ if there are infinitely many 7's in the decimal expansion of z.

The preceding examples make it clear that a function:

- Is *any* rule that assigns numbers to certain other numbers
- Doesn't have to be expressed by an algebraic formula
- Isn't necessarily a single uniform condition that applies to every number
- Isn't necessarily a rule that can be applied in practice; no one knows what rule 8 associates to π, for example—it's not been proven that every single digit from 0 to 9 occurs an unlimited number of times in π's decimal expansion (after some point, π might contain, say, only the digits 0 and 1)

- Can neglect some numbers and be unclear as to which numbers the function applies (try to determine whether the function in Example 6 applies to π, for example); the set of numbers to which a function applies, called the **domain** of the function, is discussed in the next section

Notation and Domains

Before proceeding, we need some notation. Specifically, we need a convenient way of naming functions and referring to them in general. The standard practice is to denote a function by a letter. The letter "f" is an obvious favorite, hence making "g" and "h" other favorites, but any letter or reasonable symbol is acceptable, not excluding "x" and "y", although these letters are usually reserved for indicating numbers.

If f is a function, then the number that f associates to a number x is denoted by $f(x)$. This symbol is read "f of x" and is often called the **value of f at x**. Of course, if we denote a function by x, then some other letter must be chosen to denote the number (a legitimate though depraved choice would be f, resulting in the symbol $x(f)$). Note that the symbol $f(x)$ makes sense only for x in the domain of f; for other x the symbol $f(x)$ is not defined.

If the functions defined in Examples 1–8 in the preceding section are denoted by $f, g, h, r, s, \theta, \alpha_x$, and y, then we can rewrite their definitions as

(1) $f(x) = x^2$ for all x

(2) $g(y) = \dfrac{y^3 + 3y + 5}{y^2 + 1}$ for all y

(3) $h(c) = \dfrac{c^3 + 3c + 5}{c^2 - 1}$ for all $c \neq 1, -1$

(4) $r(x) = x^2$ for all x such that $-17 \leq x \leq \pi/3$

(5) $s(x) = \begin{cases} 0, & x \text{ irrational} \\ 1, & x \text{ rational} \end{cases}$

(6) $\theta(x) = \begin{cases} 5, & x = 2 \\ \dfrac{36}{\pi}, & x = 17 \\ 28, & x = \dfrac{\pi^2}{17} \\ 28, & x = \dfrac{36}{\pi} \\ 16, & x \neq 2, 17, \dfrac{\pi^2}{17}, \dfrac{36}{\pi}, \text{ and } x = a + b\sqrt{2} \text{ for } a, b \text{ in } \mathbb{Q} \end{cases}$

(7) $\alpha_x(t) = t^2 + x$ for all numbers t

(8) $y(x) = \begin{cases} n, & \text{exactly } n \text{ 7's appear in the decimal expansion of } x \\ -\pi, & \text{infinitely many 7's appear in the decimal expansion of } x \end{cases}$

These definitions illustrate the common practice of defining a function, say f, and indicating what $f(x)$ is for every number x in the domain of f. (Note that this is exactly the same as indicating $f(a)$ for every number a, or $f(b)$ for every number b, and so on.) Certain abbreviations are acceptable in practice. Definition (1) can be written simply as

(1) $\quad f(x) = x^2$

where the qualifying phrase "for all x" is understood. The only possible abbreviation for Definition (4), however, is

(4) $\quad r(x) = x^2, \quad -17 \leq x \leq \pi/3$

It's usually understood that a definition such as

$$k(x) = \frac{1}{x} + \frac{1}{x-1}, \quad x \neq 0, 1$$

can be shortened to

$$k(x) = \frac{1}{x} + \frac{1}{x-1}$$

In other words, unless the domain is explicitly restricted further, it's understood to consist of all numbers for which the definition makes sense.

The following assertions about the functions defined above are easy to check:

$$f(x+1) = f(x) + 2x + 1$$

$$g(x) = h(x) \text{ if } x^3 + 3x + 5 = 0$$

$$r(x+1) = r(x) + 2x + 1 \text{ if } -17 \leq x \leq \frac{\pi}{3} - 1$$

$$s(x+y) = s(x) \text{ if } y \text{ is rational}$$

$$\theta\left(\frac{\pi^2}{17}\right) = \theta\left(\frac{36}{\pi}\right)$$

$$\alpha_x(x) = x \cdot [f(x) + 1]$$

$$y\left(\frac{1}{3}\right) = 0$$

$$y\left(\frac{7}{9}\right) = -\pi$$

Keep in mind that an expression like $s(a)$ is a number like any other number, so the expression $f(s(a))$ makes sense. In fact, $f(s(a)) = s(a)$ for all a (why?). After some initial disorientation, even complicated expressions aren't difficult to unravel. For example, the intimidating expression

$$f(r(s(\theta(\alpha_3(y(\frac{1}{3}))))))$$

can easily be evaluated step by step:

$$\begin{aligned}
&f(r(s(\theta(\alpha_3(y(\tfrac{1}{3})))))) \\
&= f(r(s(\theta(\alpha_3(0))))) \\
&= f(r(s(\theta(3)))) \\
&= f(r(s(16))) \\
&= f(r(1)) \\
&= f(1) \\
&= 1
\end{aligned}$$

Chapter 3 Functions

Function Types and Composition

The function defined in (1) in the preceding section is an example of an important class of functions, the polynomial functions. A function f is a **polynomial function** if there are real numbers $a_0, \ldots, a_n$ such that

$$f(x) = a_n x^n + a_{n-1} x^{n-1} + \cdots + a_2 x^2 + a_1 x + a_0 \qquad \text{for all } x$$

When $f(x)$ is written in this form, it's usually assumed that $a_n \neq 0$. The highest power of x with a nonzero coefficient is called the **degree** of f. For example, the polynomial function f defined by $f(x) = 7x^6 + 153x^4 - \pi$ has degree 6.

The functions defined in (2) and (3) in the preceding section belong to a larger class of functions called the **rational functions**, of the form p/q where p and q are polynomial functions (and q isn't a function that's always 0). The rational functions themselves are special examples of an even larger class of functions, studied deeply in calculus, which are simpler than many of the functions mentioned earlier. Some examples of rational functions are

(9) $\quad f(x) = \dfrac{x + x^2 + x \sin^2 x}{x \sin x + x \sin^2 x}$

(10) $\quad f(x) = \sin(x^2)$

(11) $\quad f(x) = \sin(\sin(x^2))$

(12) $\quad f(x) = \sin^2(\sin(\sin^2(x \sin^2 x^2))) \cdot \left(\dfrac{x + \sin(x \sin x)}{x + \sin x} \right)$

At first glance it might seem odd that the preceding functions, particularly (12), are considered to be simple, but a closer inspection reveals that they can be built up from a few simple functions by using a few simple means of combining functions. To construct the functions (9)–(12), we need to start with the identity function I, for which $I(x) = x$, and the sine function sin, whose value $\sin(x)$ at x is often written simply $\sin x$. The following examples show some of the important ways in which functions can be combined to produce new functions.

If f and g are any two functions, then we can define a new function $f+g$, called the **sum** of f and g, by the equation

$$(f+g)(x) = f(x) + g(x)$$

Note that according to the conventions we've adopted, the domain of $f+g$ consists of all x for which "$f(x) + g(x)$" makes sense; that is, the set of all x in both domain f and domain g. If A and B are any two sets, then $A \cap B$ (read "A intersect B" or "the intersection of A and B") denotes the set of x in both A and B; this notation lets us write $\mathrm{domain}(f+g) = \mathrm{domain}\, f \cap \mathrm{domain}\, g$.

Similarly, we define the **product** $f \cdot g$ and the **quotient** $\dfrac{f}{g}$ (or f/g) of f and g by

$$(f \cdot g)(x) = f(x) \cdot g(x)$$

and

$$(f/g)x = f(x)/g(x)$$

Moreover, if g is a function and c is a number, then we define a new function $c \cdot g$ by

$$(c \cdot g)(x) = c \cdot g(x)$$

This function becomes a special case of the notation $f \cdot g$ if we take the symbol c to represent the function f defined by $f(x) = c$. Such a function, which has the same value for all numbers x, is called a **constant function**.

The domain of $f \cdot g$ is domain $f \cap$ domain g, and the domain of $c \cdot g$ is simply the domain of g. The domain of f/g is somewhat complicated; it can be written as domain $f \cap$ domain $g \cap \{x : g(x) \neq 0\}$, where the symbol $\{x : g(x) \neq 0\}$ denotes the set of numbers x such that $g(x) \neq 0$. In general, $\{x : \ldots\}$ denotes the set of all x such that "..." is true. Thus $\{x : x^3 + 3 < 11\}$ denotes the set of all numbers x such that $x^3 < 8$, and consequently $\{x : x^3 + 3 < 11\} = \{x : x < 2\}$. Either of these symbols could just as well have been written using y everywhere instead of x. Variations of this notation are common, but don't merit discussion. Anyone can recognize that $\{x > 0 : x^3 < 8\}$, for example, denotes the set of positive numbers whose cube is less than 8; this symbol can be expressed more formally as $\{x : x > 0 \text{ and } x^3 < 8\}$. Incidentally, this set is equal to the set $\{x : 0 < x < 2\}$. One variation of this notation is somewhat less transparent, but standard. The set $\{1, 3, 2, 4\}$, for example, contains only the four numbers 1, 2, 3, and 4 and can also be denoted by $\{x : x = 1$ or $x = 3$ or $x = 2$ or $x = 4\}$.

Certain facts about the sum, product, and quotient of functions are obvious consequences of facts about sums, products, and quotients of numbers. For example, it's easy to prove that

$$(f + g) + h = f + (g + h)$$

The proof is characteristic of almost every proof that demonstrates that two functions are equal—the two functions are shown to have the same domain, and the same value at any number in the domain. To prove that $(f + g) + h = f + (g + h)$, for example, note that unraveling the definition of the two sides gives

$$[(f + g) + h](x) = (f + g)(x) + h(x)$$
$$= [f(x) + g(x)] + h(x)$$

and

$$[f + (g + h)](x) = f(x) + (g + h)(x)$$
$$= f(x) + [g(x) + h(x)]$$

and the equality of $[f(x) + g(x)] + h(x)$ and $f(x) + [g(x) + h(x)]$ is a fact about numbers. In this proof the equality of the two domains wasn't explicitly mentioned because it's immediately obvious for these equations—the domain of $(f + g) + h$ and of $f + (g + h)$ is clearly domain $f \cap$ domain $g \cap$ domain h. We naturally write $f + g + h$ for $(f + g) + h = f + (g + h)$, just as we do for numbers.

It's also easy to prove that $(f \cdot g) \cdot h = f \cdot (g \cdot h)$, and this function is denoted by $f \cdot g \cdot h$. Proving that the equations $f + g = g + f$ and $f \cdot g = g \cdot f$ proceeds along similar lines.

Using the operations $+$, $\cdot$, and $/$ we can now express the function f defined in (9) by

(9) $\quad f = \dfrac{I + I \cdot I + I \cdot \sin \cdot \sin}{I \cdot \sin + I \cdot \sin \cdot \sin}$

It should be clear, however, that we can't express function (10) this way. We need yet another way of combining functions. This combination, the composition of two functions, is by far the most important.

If f and g are any two functions, then we define a new function $f \circ g$, called the **composition** of f and g, by

$$(f \circ g)(x) = f(g(x))$$

The domain of $f \circ g$ is $\{x : x$ is in domain g and $g(x)$ is in domain $f\}$. The symbol "$f \circ g$" is often read "f circle g". The alternative phrase "the composition of f and g" is more cumbersome and carries the danger of confusing $f \circ g$ with $g \circ f$. These two expressions mustn't be confused because they're *not* usually equal. In fact, almost any f and g chosen at random will illustrate this point (try $f = I \cdot I$ and $g = \sin$, for example). Happily, however, composition *is* associative:

$$(f \circ g) \circ h = f \circ (g \circ h)$$

The proof is trivial, and this function is denoted by $f \circ g \circ h$. We can now write the functions (10), (11), and (12) as

(10) $f = \sin \circ (I \cdot I)$

(11) $f = \sin \circ \sin \circ (I \cdot I)$

(12) $f = (\sin \cdot \sin) \circ \sin \circ (\sin \cdot \sin) \circ (I \cdot [(\sin \cdot \sin) \circ (I \cdot I)]) \cdot$

$$\sin \circ \left(\frac{I + \sin \circ (I \cdot \sin)}{I + \sin} \right)$$

Although this method of writing functions reveals their "structure" clearly, it's neither short nor convenient. The shortest unambiguous name for the function f such that $f(x) = \sin(x^2)$ for all x is unfortunately the same cumbersome expression: "the function f such that $f(x) = \sin(x^2)$ for all x". Despite the passing of centuries, mathematicians have yet to agree on how to abbreviate this description. A strong contender is currently

$$x \to \sin(x^2)$$

This expression is read "x goes to $\sin(x^2)$" or "x arrow $\sin(x^2)$", but this notation is rarely used in mainstream precalculus or calculus textbooks, where instead it's common to see the abbreviation "the function $f(x) = \sin(x^2)$" or the dreadful "the function $\sin(x^2)$". The latter abbreviation so lacks precision that it actually confuses a number and a function, so should be avoided, especially in print (though I admit that I've adopted it for personal use, as many people do whether they want to or not). Conventions and abbreviations are useful provided that any slight logical deficiencies cause no confusion, but use a precise description if there's a chance of confusion. For example, the ambiguous phrase "the function $x + t^3$" could mean either

$$x \to x + t^3 \text{ (that is, the function } f \text{ such that } f(x) = x + t^3 \text{ for all } x)$$

or

$$t \to x + t^3 \text{ (that is, the function } f \text{ such that } f(t) = x + t^3 \text{ for all } t)$$

As you progress in mathematical sophistication to analysis and beyond, you'll see that many important concepts associated with functions have "$x \to$" built in to the notation.

Formal Definitions

We've arrived at the point where we can reconsider our provisional definition of a function, which you'll recall is:

> **Provisional Definition** A **function** is a rule that assigns, to each of certain real numbers, some other real number.

By now the word "rule" probably strikes you as unclear, and you might be asking a perfectly legitimate question like "What happens if you break this rule?" Another objection to the use of the word "rule" is that

$$f(x) = x^2$$

and

$$f(x) = x^2 + 3x + 3 - 3(x + 1)$$

are certainly *different* rules, if by a rule we mean the actual instructions given for determining $f(x)$. Nevertheless, we want

$$f(x) = x^2$$

74   Essential Advanced Precalculus

and

$$f(x) = x^2 + 3x + 3 - c(x + 1)$$

to define the same function. For this reason, a function is sometimes defined as an "association" between numbers. Unfortunately, "association" is even more vague than "rule".

A satisfactory way to define functions can't be constructed by finding synonyms for troublesome words. The formal definition that mathematicians use for "function" blends intuitive ideas and rigorous mathematics. The correct question to ask about a function isn't "What is a rule?" or "What is an association?" but "What does one have to know about a function to know all about it?" The answer to this last question is easy: for each number x one needs to know the number $f(x)$. Imagine a table displaying all the information that we could desire about the function $f(x) = x^2$:

x	$f(x)$
1	1
-1	1
2	4
-2	4
$\sqrt{2}$	2
$-\sqrt{2}$	2
π	π^2
$-\pi$	π^2

It's even unnecessary to arrange the numbers in the table (which would nevertheless be impossible if we wanted to list them all). Instead of using a two-column array, we can consider various pairs of numbers simply collected together into a set:

$$(1, 1), (-1, 1), (2, 4), (-2, 4), (\pi, \pi^2), (\sqrt{2}, 2), \ldots$$

The pairs here are called **ordered pairs**, to emphasize that, for example, (2, 4) is not the same pair as (4, 2). We're going to define functions in terms of ordered pairs but leave "ordered pair" undefined until the next section.

To find $f(1)$ we simply take the second number of the pair whose first member is 1. To find $f(\pi)$ we take the second number of the pair whose first member is π. We're saying that a function can just as well be defined as a collection of pairs of numbers. For example, if we were given the following collection (which contains only five pairs):

$$f = \{(1, 7), (3, 7), (5, 3), (4, 8), (8, 4)\}$$

then $f(1) = 7, f(3) = 7, f(5) = 3, f(4) = 8, f(8) = 4$, and 1, 3, 4, 5, 8 are the only numbers in the domain of f. If we consider the collection

$$f = \{(1, 7), (3, 7), (2, 5), (1, 8), (8, 4)\}$$

then $f(3) = 7, f(2) = 5, f(8) = 4$; but it's impossible to decide whether $f(1) = 7$ or $f(1) = 8$. In other words, a function can't be defined to be just any collection of pairs of numbers, so we'll rule out the possibility of ambiguity that arose in the preceding example. We therefore arrive at the following rigorous definition:

> **Definition** A **function** is a collection of pairs of numbers with the following property: if (a, b) and (a, c) are both in the collection, then $b = c$; in other words, the collection must not contain two different pairs with the same first element.

Rigorous definitions in mathematics are at least as important as theorems, and it's crucial to distinguish them from comments, motivating remarks, and casual explanations.

One more definition (actually defining two things simultaneously) can now be made rigorously:

> **Definition** If f is a function, then the **domain** of f is the set of all a for which there is some b such that (a, b) is in f. If a is in the domain of f, then it follows from the definition of a function that there is, in fact, a unique number b such that (a, b) is in f. This unique b is denoted by $f(a)$.

With this definition we've reached our goal of showing that the important characteristic of a function f is that a number $f(x)$ is determined for each number x in its domain. You might now feel that your cozy intuitive definition has been stomped by a formal abstraction. Take comfort in two consolations. First, although a function is defined to be a collection of pairs, there's nothing to stop you from thinking of a function as a rule. Second, neither the intuitive nor the formal definition is the best way of thinking about functions—the best way is to draw graphs (Chapter 4).

Ordered Pairs

The concept of an ordered pair of objects is necessary in the definition of a function. The preceding section made use of ordered pairs without defining them or even explicitly stating their properties. The only property that needs to be stated formally is that an ordered pair (a, b) is determined by a and b and the order in which they're given:

$$\text{If } (a, b) = (c, d), \text{ then } a = c \text{ and } b = d.$$

We can treat ordered pairs conveniently by introducing (a, b) as an undefined term and then regarding the basic property as axiomatic. Because this property is the only significant fact about ordered pairs, there's little point worrying about what an ordered pair "truly" is. But if simply defining this basic property to be a theorem leaves you uneasy, then read on.

We needn't restrict ourselves to ordered pairs of numbers. It's reasonable and important to consider ordered pairs of any two mathematical objects, meaning we must modify our definition to involve only concepts common to all areas of mathematics. The only omnipresent mathematical concept is the set, and ordered pairs (like everything else in mathematics) can be defined in terms of sets—a special sort of set in this case.

Sadly, the obvious first choice of the set $\{a, b\}$, containing the two elements a and b, won't work as a definition for (a, b) because there's no way to determine from $\{a, b\}$ which of a or b is meant to be the first element.

It turns out, surprisingly, that the set we're looking for is

$$\{\{a\}, \{a, b\}\}$$

This set has two members, both of which are *themselves* sets: one member is the set {a}, containing the single element a, and the other is the set {a, b}. Now, define (a, b) to be this set:

Definition $(a, b) = \{\{a\}, \{a, b\}\}$

The justification for this choice is the theorem already given:

Theorem 3.1   If $(a, b) = (c, d)$, then $a = c$ and $b = d$.

Proof   By definition

$$\{\{a\}, \{a, b\}\} = \{\{c\}, \{c, d\}\}$$

Now $\{\{a\}, \{a, b\}\}$ contains only two members: $\{a\}$ and $\{a, b\}$; and a is the only common element of these two members of $\{\{a\}, \{a, b\}\}$. Similarly, c is the unique common element of both members of $\{\{c\}, \{c, d\}\}$. Therefore $a = c$. Hence we have

$$\{\{a\}, \{a, b\}\} = \{\{a\}, \{a, d\}\}$$

and only the proof that $b = d$ remains. It's convenient to distinguish two cases:

Case 1. $b = a$. In this case, $\{a, b\} = \{a\}$, so the set $\{\{a\}, \{a, b\}\}$ actually has only one member: $\{a\}$. The same must be true of $\{\{a\}, \{a, d\}\}$, so $\{a, d\} = \{a\}$, which implies that $d = a = b$.

Case 2. $b \ne a$. In this case, b is in one member of $\{\{a\}, \{a, b\}\}$ but not in the other. It must therefore be true that b is in one member of $\{\{a\}, \{a, d\}\}$ but not in the other. This can happen only if b is in $\{a, d\}$, but b is not in $\{a\}$; thus $b = a$ or $b = d$, but $b \ne a$, so $b = d$. ∎

Problems

1. Let $f(x) = 1/(1 + x)$. What is

   (a) $f(f(x))$ (for which x does this expression make sense?)

   (b) $f(1/x)$

   (c) $f(cx)$

   (d) $f(x + y)$

(e) $f(x) + f(y)$

(f) For which numbers c is there a number x such that $f(cx) = f(x)$. (Hint: There are many more than you might think at first glance.)

(g) For which numbers c is it true that $f(cx) = f(x)$ for two different numbers x?

2. Let $g(x) = x^2$, and let $h(x) = \begin{cases} 0, & x \text{ irrational} \\ 1, & x \text{ rational} \end{cases}$.

   (a) For which y is $h(y) \leq y$?
   (b) For which y is $h(y) \leq g(y)$?
   (c) What is $g(h(z)) - h(z)$?
   (d) For which w is $g(w) \leq w$?
   (e) For which v is $g(g(v)) = g(v)$?

3. Find the domain of the functions defined by the following formulas.

   (a) $f(x) = \sqrt{1-x^2}$
   (b) $f(x) = \sqrt{1-\sqrt{1-x^2}}$
   (c) $f(x) = \dfrac{1}{x-1} + \dfrac{1}{x-2}$
   (d) $f(x) = \sqrt{1-x^2} + \sqrt{x^2-1}$
   (e) $f(x) = \sqrt{1-x} + \sqrt{x-2}$

4. Let $S(x) = x^2$, let $P(x) = 2^x$, and let $s(x) = \sin x$. Find each of the following. In each case your answer must be a *number*.
   (a) $(S \circ P)(y)$
   (b) $(S \circ s)(y)$
   (c) $(S \circ P \circ s)(t) + (s \circ P)(t)$
   (d) $s(t^3)$

5. Express each of the following functions in terms of S, P, and s (given in the preceding problem), using only $+$, $\cdot$, and $\circ$ (the answer to part (a) is $P \circ s$, for example). In each case your answer must be a *function*.
   (a) $f(x) = 2^{\sin x}$
   (b) $f(x) = \sin 2^x$

(c) $f(x) = \sin x^2$

(d) $f(x) = \sin^2 x$

Note: Remember that $\sin^2 x$ is an abbreviation for $(\sin x)^2$.

(e) $f(t) = 2^{2^t}$

Note: a^{b^c} always means $a^{(b^c)}$; this convention is adopted because $(a^b)^c$ can be written more simply as a^{bc}.

(f) $f(u) = \sin(2^u + 2^{u^2})$

(g) $f(y) = \sin(\sin(\sin(2^{2^{2^{\sin y}}})))$

(h) $f(a) = 2^{\sin^2 a} + \sin(a^2) + 2^{\sin(a^2 + \sin a)}$

6. For which numbers a, b, c, and d will the function
$$f(x) = \frac{ax+b}{cd+d}$$
satisfy $f(f(x)) = x$ for all x?

7. (a) If A is any set of real numbers, define a function C_A as follows:
$$C_A(x) = \begin{cases} 0, & x \text{ in } A \\ 1, & x \text{ not in } A \end{cases}$$
Find expressions for $C_{A \cap B}$ and $C_{A \cup B}$ and $C_{\mathbb{R}-A}$, in terms of C_A and C_B.

The symbol $A \cap B$ was defined in "Function Types and Composition". The symbols $A \cup B$ (read "A union B" or "the union of A and B") and $\mathbb{R} - A$ (read "$\mathbb{R}$ difference A" or "the difference of $\mathbb{R}$ and A") are defined as follows:

$A \cup B = \{x : x \text{ is in } A \text{ or } x \text{ is in } B\}$
$\mathbb{R} - A = \{x : x \text{ is in } \mathbb{R} \text{ but } x \text{ is not in } A\}$

where $\mathbb{R}$ is the set of all real numbers (rational and irrational).

(b) Suppose that f is a function such that $f(x) = 0$ or 1 for each x. Prove that there is a set A such that $f = C_A$.

(c) Show that $f = f^2$ if and only if $f = C_A$ for some set A.

8. (a) For which functions f is there a function g such that $f = g^2$?
   (Hint: First answer the question by mentally replacing "function" by "number".)

   (b) For which functions f is there a function g such that $f = 1/g$?

9. (a) Suppose that H is a function and y is a number such that $H(H(y)) = y$. What is:
$$\underbrace{H(H(H(\cdots(H(y))\cdots)}_{80 \text{ times}}$$

   (b) Answer part (a) if 80 is replaced by 81.

   (c) Answer part (a) if $H(H(y)) = H(y)$.

10. A function f is **even** if $f(x) = f(-x)$ and **odd** if $f(x) = -f(-x)$. For example, f is even if $f(x) = x^2$ or $f(x) = |x|$ or $f(x) = \cos x$, while f is odd if $f(x) = x$ or $f(x) = \sin x$.

    (a) Determine whether $f + g$ is even, odd, or not necessarily either, in the four cases obtained by choosing f even or odd, and g even or odd. Display your answers in a 2 × 2 table.

    (b) Do the same for $f \cdot g$.

    (c) Do the same for $f \circ g$.

11. Prove or give a counterexample for each of the following assertions.

    (a) $f \circ (g + h) = f \circ g + f \circ h$

    (b) $(g + h) \circ f = g \circ f + h \circ f$

    (c) $\dfrac{1}{f \circ g} = \dfrac{1}{f} \circ g$

    (d) $\dfrac{1}{f \circ g} = f \circ \dfrac{1}{g}$

12. Suppose that $g = h \circ f$. Prove that if $f(x) = f(y)$, then $g(x) = g(y)$.

13. Suppose that $f \circ g = I$, where $I(x) = x$.
    (a) Prove by contradiction that if $x \neq y$, then $g(x) \neq g(y)$.
    (b) Prove that every number b can be written $b = f(a)$ for some number a.

14. Suppose that $f(x) = x + 1$. Are there any functions g such that $f \circ g = g \circ f$?

15. Suppose that f is a constant function. For which functions g does $f \circ g = g \circ f$?

16. Suppose that $f \circ g = g \circ f$ for *all* functions g. Show that f is the identity function, $f(x) = x$.

17. Suppose that we define $f < g$ to mean that $f(x) < g(x)$ for all x.
    (a) If $f < g$, then is $h \circ f < h \circ g$?
    (b) Is $f \circ h < g \circ h$?

4

Graphs

The Real Line and Intervals

At the mention of the real numbers, an image of a straight line will almost certainly form in the mind of any mathematician or mathematics student. This mental picture of the real numbers—which should be neither casually dismissed nor crushingly embraced—provides a geometric intuition for interpreting statements about numbers, and can even suggest ways to prove them.

You're probably familiar with the conventional method of using a straight line to picture the real numbers. Imagine a horizontal straight line, extending endlessly in both directions. Choose an arbitrary point, called the **origin** or **zero point**, and label it 0. Pick another arbitrary point to the right of 0 and label it 1. The distance between these two points (the unit distance) is the measuring scale that associates every real number with a point on the line.

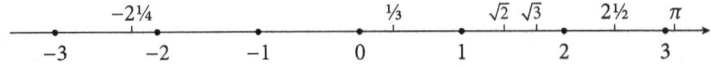

With this arrangement, if $a < b$, then the point corresponding to a lies to the left of the point corresponding to b. (The arrowhead on the right end of the real line indicates the positive direction.) You can draw rational numbers, such as ½, in the obvious way. The irrational numbers also fit into this scheme, so that every real number can be drawn as a point on the line. You can plot irrational numbers on the line by using their decimal expansions: $\sqrt{2} = 1.414\ldots$, $\sqrt{3} = 1.732\ldots$, and

$\pi = 3.14159\ldots$. (The decimal expansion of an irrational number never repeats or terminates, unlike a rational number.) We won't quibble about justifying this assumption because this method of "drawing" numbers is solely a pleasing method of picturing abstract ideas. Our proofs will never rely on pictures, although such pictures frequently suggest or help explain proofs. Because this geometric picture plays a prominent (if inessential) role, geometric terminology is often used when speaking of numbers—thus a number is sometimes called a **point**, and $\mathbb{R}$ is often called the **real line** or the **number line**. (By convention, the set of all real numbers—that is, rational and irrational numbers—is denoted by the symbol $\mathbb{R}$.)

The number $|a - b|$ has a simple interpretation in terms of this geometric picture: it's the distance between a and b (that is, the length of the line segment that has a as one end point and b as the other). This means, to choose a common example, that the set of numbers x which satisfy $|x - a| < \varepsilon$ can be pictured as the collection of points whose distance from a is less than ε. This set of points is the **interval** from $a - \varepsilon$ to $a + \varepsilon$, which can also be described as the points corresponding to numbers x with $a - \varepsilon < x < a + \varepsilon$, as shown in the following figure. (In mathematics, an arbitrarily small positive quantity is commonly denoted by ε, the lowercase Greek letter epsilon.)

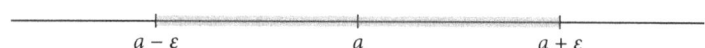

Sets of numbers that correspond to intervals arise so frequently that they have special names. The set $\{x : a < x < b\}$ is denoted by (a, b) and called the **open interval** from a to b. This notation can create ambiguity because (a, b) is also used to denote a pair of numbers, but a careful writer will always make it clear whether (a, b) refers to an interval or a pair. Note that if $a \geq b$, then $(a, b) = \emptyset$, the set with no elements; in practice, however, it's almost always assumed (explicitly or implicitly) that the number a is less than b in an interval (a, b).

The set $\{x : a \leq x \leq b\}$ is denoted by $[a, b]$ and is called the **closed interval** from a to b. This symbol is usually reserved for the case $a < b$, but it's sometimes also used for $a = b$.

The following figure shows how the intervals (a, b) and $[a, b]$ are usually depicted. Note the pictorial conventions and keep in mind that no reasonably accurate picture could ever depict the actual difference between the intervals (a, b) and $[a, b]$.

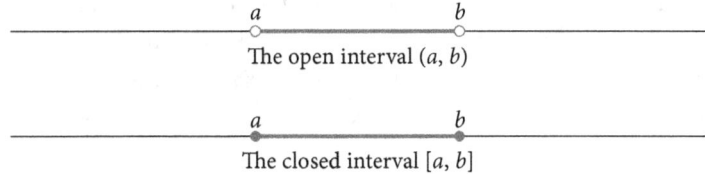

The following figure shows certain infinite intervals. The set $\{x : x > a\}$ is denoted by (a, ∞), and the set $\{x : x \geq a\}$ is denoted by $[a, \infty)$. The sets $(-\infty, a)$ and $(-\infty, a]$ are defined similarly. The set $\mathbb{R}$ of all real numbers is also considered to be an interval, and is sometimes denoted by $(-\infty, \infty)$.

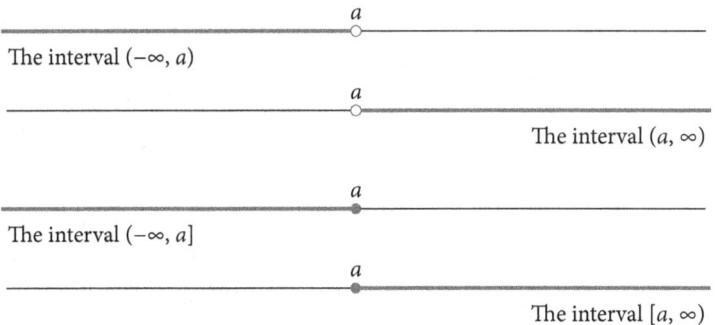

Heed the standard warning: the symbols ∞ and $-\infty$, though usually read "infinity" and "minus infinity", are *purely* suggestive. There's no number ∞ that satisfies $\infty \geq a$ for all numbers a. Though the symbols ∞ and $-\infty$ appear in many contexts, it's always necessary to define them in ways that refer to only numbers.

Coordinates, Slope, and Distance

Of greater interest to us than the method of drawing numbers is a method of drawing pairs of numbers. We can extend the method used to build the real line to assign coordinates to points in a plane. This procedure, probably also familiar to you, requires a **coordinate system** specified by two straight lines intersecting at right angles. To distinguish these straight lines, we call one the **horizontal axis** and the other the **vertical axis**. (Calling these lines the "first" and "second" axes is logically preferable, but the more-descriptive terms "horizontal" and "vertical" have stuck.) Each of the two axes can be labeled with real numbers, but we can also label points on the horizontal axis with pairs $(a, 0)$ and points on the vertical axis with pairs $(0, b)$, so that the intersection of the two axes, called the **origin** of the coordinate system, is labeled $(0, 0)$. Any pair (a, b) can now be drawn as shown in the accompanying figure, lying at the vertex of the rectangle whose other three vertices are labeled $(0, 0)$, $(a, 0)$, and $(0, b)$. The numbers a and b are called the **first coordinate** and **second coordinate**, respectively, of a point (a, b) determined in this way.

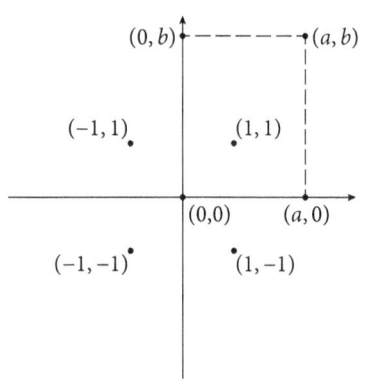

Our real concern in this book is a method of drawing functions. Because a function is just a collection of pairs of numbers, we can draw a function by drawing each of the pairs in the function. The drawing obtained in this way is called the **graph** of the function. In other words, the graph of f contains all the points corresponding to pairs $(x, f(x))$. Even though most functions contain infinitely many pairs, many functions have graphs that are quite easy to draw.

Unsurprisingly, the simplest functions of all, the constant functions $f(x) = c$, have the simplest graphs. It's easy to see that the graph of the function $f(x) = c$ is a straight line parallel to the horizontal axis, at distance c from it, as shown in the following figure.

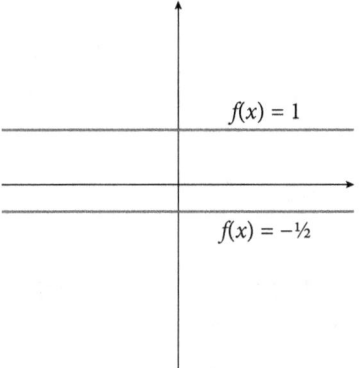

The functions $f(x) = cx$ also have simple graphs—straight lines through $(0, 0)$, as shown in the following figure.

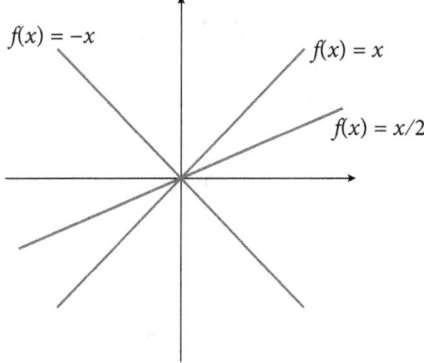

Refer to the following figure for a proof of the preceding fact: let x be some number not equal to 0, and let L be the straight line that passes through the origin O, corresponding to $(0, 0)$, and through the point A, corresponding to (x, cx).

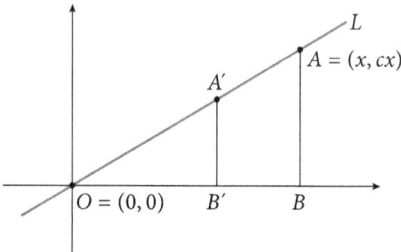

Now, a point A', with first coordinate y, will lie on L when the triangle $A'B'O$ is similar to the triangle ABO—that is, when

$$\frac{A'B'}{OB'} = \frac{AB}{OB} = c$$

This condition is precisely where A' corresponds to the pair (y, cy)—that is, where A' lies on the graph of f. This argument has implicitly assumed that $c > 0$, but the other cases can be demonstrated easily. The number c, which measures the ratio of the sides of the triangles appearing in the proof, is called the **slope** of the straight line, and any line parallel to this line is also said to have slope c.

The preceding demonstration isn't a *formal* proof—a rigorous demonstration would require methods that are well beyond the scope of this book. In fact, a rigorous proof of *any* statement that connects geometric and algebraic concepts would first require a real proof (or a precisely stated assumption) that the points on a straight line correspond in an exact way to the real numbers. Additionally, it would be necessary to develop plane geometry and the properties of real numbers precisely. The detailed development of plane geometry is a beautiful subject, but it's not a prerequisite for the study of precalculus or calculus. For our purposes, simple definitions will suffice. We define the plane to be the set of all pairs of real numbers, and we define straight lines to be certain collections of pairs, including, among others, the collections $\{(x, cx) : x$ is a real number$\}$. We'll use geometric pictures only as an aid to intuition.

To provide this artificially constructed geometry with all the structure of geometry studied in high school, one more definition is required. If (a, b) and (c, d) are two points in the plane—that is, pairs of real numbers—then we define the **distance** between (a, b) and (c, d) to be

$$\sqrt{(a-c)^2 + (b-d)^2}$$

With this definition, the Pythagorean theorem is now part of our geometry, as illustrated in the following figure.

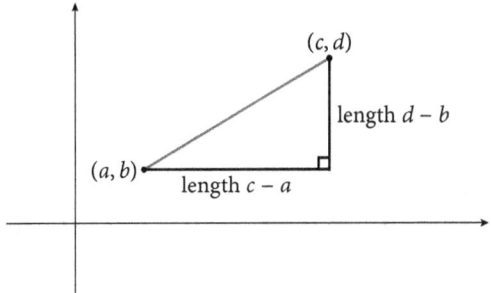

Linear Functions

By using the informal geometric picture sketched in the preceding two sections, it's easy to see that the graph of the function $f(x) = cx + d$ is a straight line with slope c, passing through the point $(0, d)$, as shown in the following figure. For this reason, the functions $f(x) = cx + d$ are called **linear functions**.

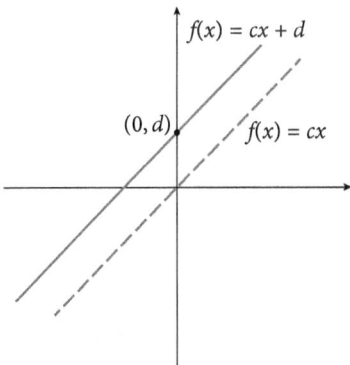

Simple as they are, linear functions occur frequently, so you should feel comfortable working with them. The following problem is typical of those involving linear functions. Given two distinct points (a, b) and (c, d), find the linear function f whose graph goes through (a, b) and (c, d). This amounts to saying that $f(a) = b$ and $f(c) = d$. If f is to be of the form $f(x) = \alpha x + \beta$, then we must have

$$\alpha a + \beta = b$$
$$\alpha c + \beta = d$$

Therefore $\alpha = (d - b)/(c - a)$ and $\beta = b - [(d - b)/(c - a)]a$, so

$$f(x) = \frac{d-b}{c-a}x + b - \frac{d-b}{c-a}a = \frac{d-b}{c-a}(x-a) + b$$

This formula is most easily remembered by using the **point-slope form** (see Problem 6).

Of course, this solution is possible only if $a \neq c$; the graphs of linear functions encompass only straight lines which aren't parallel to the vertical axis. The vertical straight lines aren't the graph of *any* function—in fact, the graph of a function can never contain even two distinct points on the same vertical line. This conclusion is immediate from the definition of a function—two points on the same vertical line correspond to pairs of the form (a, b) and (a, c) and, by definition, a function can't contain (a, b) and (a, c) if $b \neq c$. Conversely, if a set of points in the plane has the property that no two points lie on the same vertical line, then it's certainly the graph of a function. Hence, the two sets in the following figure *aren't* graphs of functions.

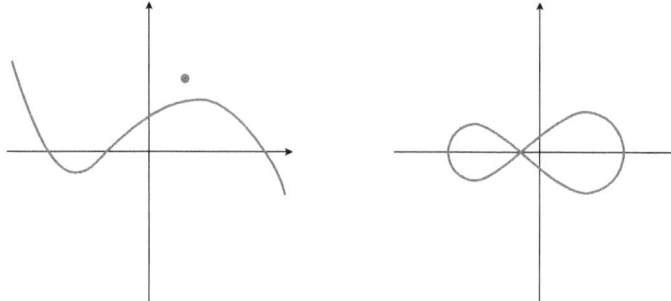

The two sets in the following figure *are* graphs of functions. Note that the example on the right is the graph of a function whose domain is not all of ℝ, since some vertical lines have no points on them.

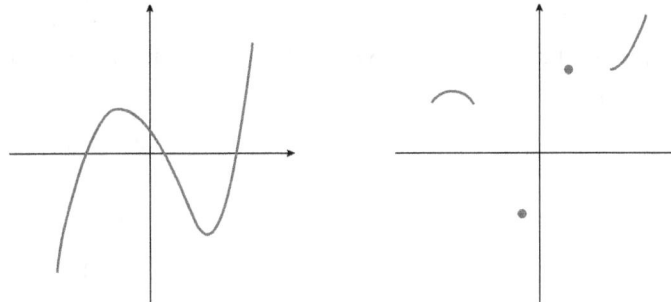

Parabolas and Power Functions

After the linear functions the simplest is perhaps the function $f(x) = x^2$. If we draw some of the pairs in f—that is, some of the pairs of the form (x, x^2)—then we get a picture like the following figure.

(−2½, 6¼) •   • (2½, 6¼)

(−2, 4) •   • (2, 4)

(−1, 1) •   • (1, 1)
            (0, 0)

It's not hard to convince yourself that all the pairs (x, x^2) lie along a curve like the one shown in the following figure. This curve is called a **parabola**.

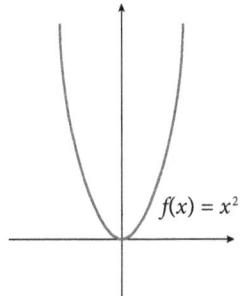

$f(x) = x^2$

Chapter 4 Graphs 91

Because a graph is only ink on paper (or pixels on a screen), the question "Is this what the graph *really* looks like?" is hard to phrase sensibly. No mathematical drawing is ever really correct. Nevertheless, there are some questions that you *can* ask: for example, how can you be sure that the graph doesn't look like one of the drawings in the following figure?

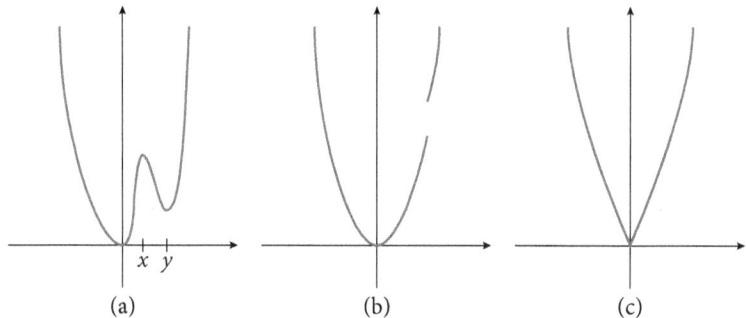

It's easy to see (and even to prove) that the graph can't look like (a); for if $0 < x < y$, then $x^2 < y^2$, so the graph must be higher at y than at x, which isn't the case in (a). It's also easy to see, simply by drawing a graph accurately, first plotting many pairs (x, x^2), that the graph can't have a large "jump" as in (b) or a "corner" as in (c). To prove these assertions, however, we first need to state mathematically what it means for a function not to have a "jump" or "corner". These ideas involve some of the fundamental concepts of calculus. As you progress mathematically, you'll be able to define them rigorously.

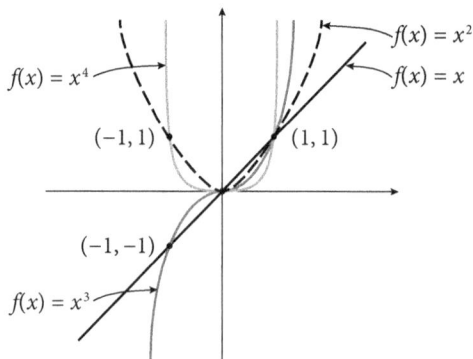

The functions $f(x) = x^n$, for various natural numbers n, are called **power functions**. Their graphs are most easily compared by superimposing several of them, as shown in the accompanying figure.

Polynomial Functions

The power functions are only special cases of polynomial functions. A **polynomial**, or more precisely, a **polynomial in x**, is an algebraic expression that's built up from the variable x and any constants by means of addition, subtraction, and multiplication alone. Two particular polynomial functions are graphed in the following figure.

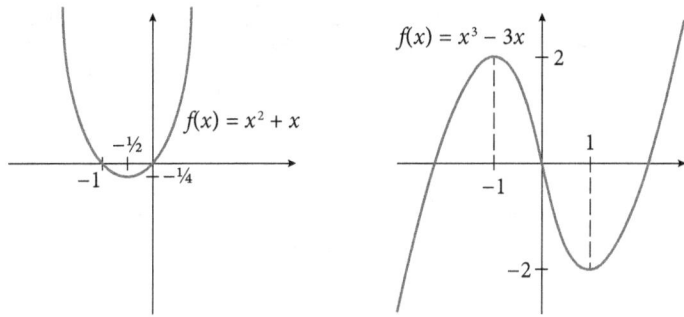

The graphs in the following figure are meant to give you a general idea of the graph of the polynomial function

$$f(x) = a_n x^n + a_{n-1} x^{n-1} + \cdots + a_2 x^2 + a_1 x + a_0$$

in the case $a_n > 0$.

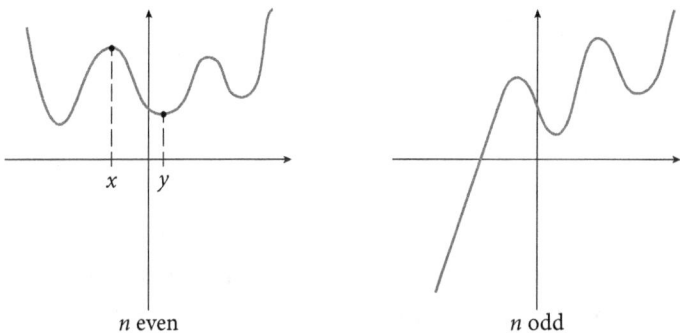

In general, the graph of f will have at most $n - 1$ "peaks" or "valleys" (a "peak" is a point like $(x, f(x))$ in the preceding figure, whereas a "valley" is a point like $(y, f(y))$. The number of peaks and valleys might actually be much smaller (the power functions, for example, have at most one valley). Although these assertions are easy to make and prove, the proofs require the powerful tools of calculus.

Rational Functions

Just as rational numbers are quotients of integers, rational functions are quotients of polynomials. A **rational function** takes the form p/q where p and q are polynomial functions (and q isn't a function that's always 0). The following figures illustrate the graphs of several rational functions, which exhibit even greater variety than do the polynomial functions.

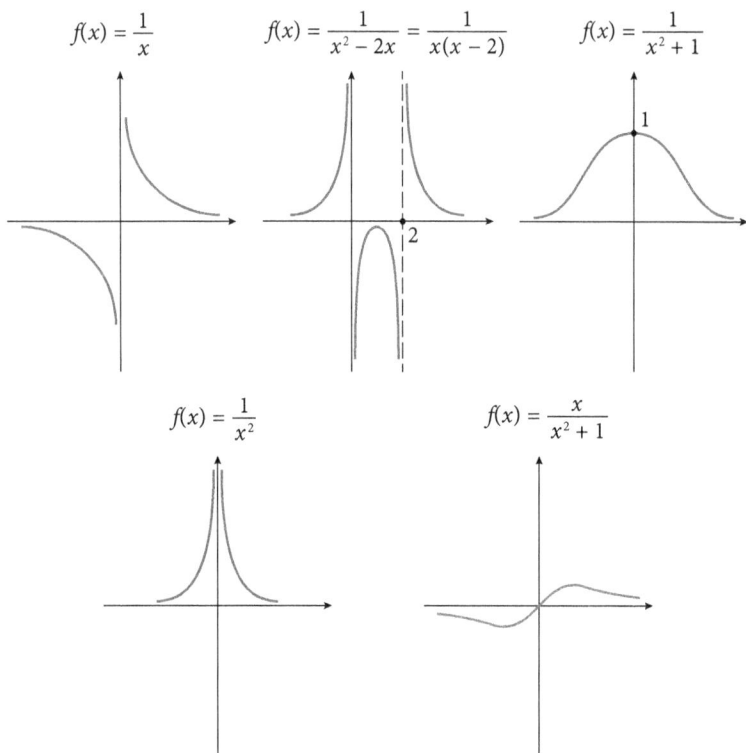

$f(x) = \dfrac{1}{x}$ $\qquad f(x) = \dfrac{1}{x^2 - 2x} = \dfrac{1}{x(x-2)} \qquad f(x) = \dfrac{1}{x^2 + 1}$

$f(x) = \dfrac{1}{x^2} \qquad\qquad f(x) = \dfrac{x}{x^2 + 1}$

Examine these graphs, paying special attention to the zeros of the denominator of a rational function. The best method for sketching a graph is to learn the characteristic features of various functions (zeros, turning points, and so on) and use these features as the basis for the sketch. (As a last resort, you can plot a few points and connect them.)

Oscillating Functions

Many interesting graphs can be constructed by "piecing together" the graphs of functions that we've already seen. The graph in the following figure is made up entirely of straight lines.

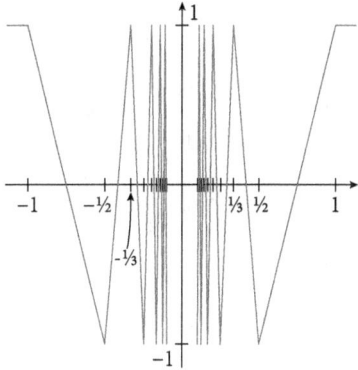

The function f with this graph satisfies

$$f\left(\frac{1}{n}\right)=(-1)^{n+1}, \quad f\left(\frac{-1}{n}\right)=(-1)^{n+1}, \quad f(x)=1, \quad |x|\geq 1$$

and is a linear function on each interval $[1/(n+1), 1/n]$ and $[-1/n, -1/(n+1)]$. (The number 0 isn't in the domain of f.) Of course, you could write out an explicit formula for $f(x)$ when x is in $[1/(n+1), 1/n]$—an exercise in the use of linear functions that would convince you that a picture is worth a thousand words.

It's actually possible to use the sine function to define, in a much simpler way, a function that exhibits this same property of oscillating infinitely often near 0. For our purposes, it's easiest to use degrees for measuring angles rather than the usual radians. The graph of the sine function is shown in the following figure.

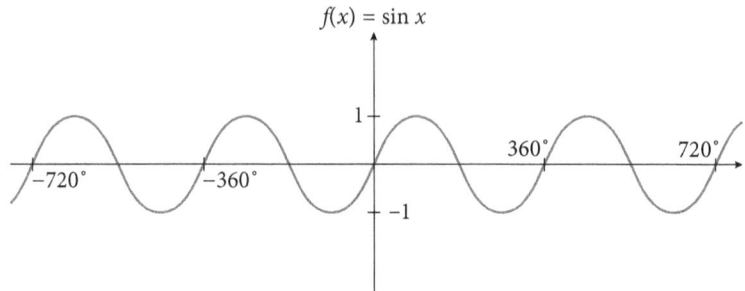

$f(x) = \sin x$

Now consider the function $f(x) = \sin 1/x$. The graph of f is shown in the following figure.

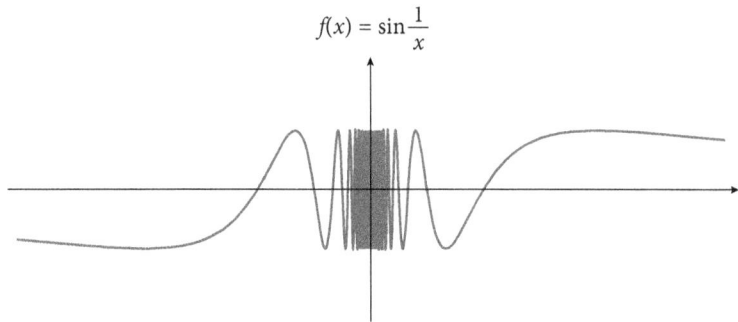

$f(x) = \sin \dfrac{1}{x}$

To draw the graph of $f(x) = \sin 1/x$, it helps to first observe that

$f(x) = 0$ for $x = \dfrac{1}{180}, \dfrac{1}{360}, \dfrac{1}{540}, \ldots,$

$f(x) = 1$ for $x = \dfrac{1}{90}, \dfrac{1}{90+360}, \dfrac{1}{90+720}, \ldots,$

$f(x) = -1$ for $x = \dfrac{1}{270}, \dfrac{1}{270+360}, \dfrac{1}{270+720}, \ldots,$

Notice that when x is large, so that $1/x$ is small, $f(x)$ is also small; when x is "large negative"—that is, when $|x|$ is large for negative x—again $f(x)$ is close to 0, although $f(x) < 0$.

An interesting modification of this function is $f(x) = x \sin 1/x$. The graph of this function is sketched in the following figure.

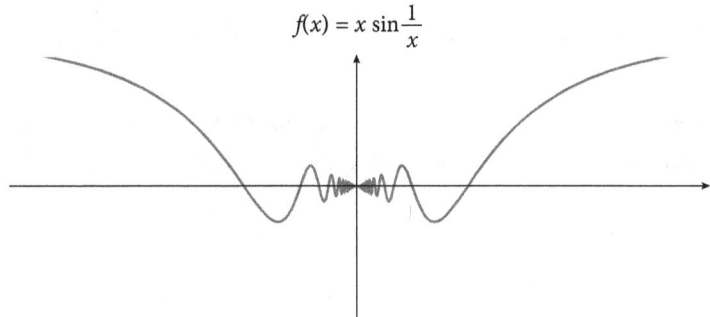

$$f(x) = x \sin \frac{1}{x}$$

Because $\sin 1/x$ oscillates infinitely often near 0 between 1 and −1, the function $f(x) = x \sin 1/x$ oscillates infinitely often between x and $-x$. The behavior of the graph for x large or large negative is harder to analyze. Because $\sin 1/x$ is getting close to 0, while x is getting larger and larger, there seems to be no telling what the product will do. It *is* possible to decide, but this question is best deferred until calculus. For comparison, the graph of $f(x) = x^2 \sin 1/x$ is sketched in the next figure.

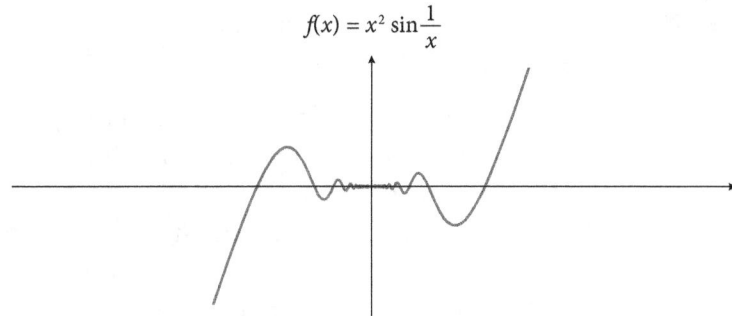

$$f(x) = x^2 \sin \frac{1}{x}$$

For these infinitely oscillating functions, it's clear that the graph can't hope to be very accurate. The best we can do is to show part of it, and leave out the part near 0 (which is the interesting part). Actually, it's easy to find much simpler functions whose graphs can't be drawn with much accuracy. The graphs of

$$f(x) = \begin{cases} x^2, & x < 1 \\ 2, & x \geq 1 \end{cases} \quad \text{and} \quad g(x) = \begin{cases} x^2, & x \leq 1 \\ 2, & x > 1 \end{cases}$$

can be distinguished only by some convention similar to that used for open and closed intervals, as shown in the following figure.

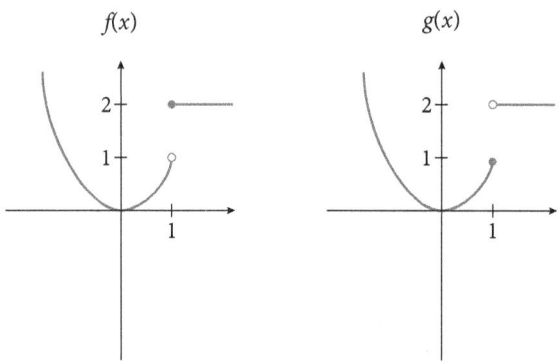

Our last example is a function whose graph is especially undrawable:

$$f(x) = \begin{cases} 0, & x \text{ irrational} \\ 1, & x \text{ rational} \end{cases}$$

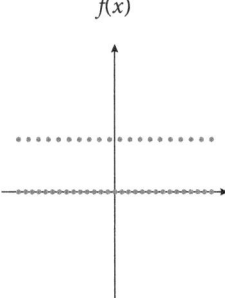

The graph of f must contain infinitely many points on the horizontal axis and also infinitely many points on a line parallel to the horizontal axis, but it mustn't contain either of these lines entirely. The accompanying figure shows the usual textbook picture of the graph. To distinguish the two parts of the graph, the dots are placed closer together on the line corresponding to irrational x. (The mathematical reason behind this convention relies on some sophisticated ideas.)

Ellipses and Hyperbolas

The idiosyncrasies of some functions in the preceding sections make it easy to forget some of the simplest and most important subsets of the plane, which aren't the graphs of functions. The most important example is the **circle**. A circle with center (a, b) and radius $r > 0$ contains, by definition, all the points (x, y) whose distance from (a, b) is equal to r, as shown in the following figure.

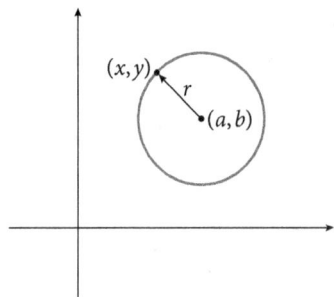

The circle thus consists of all points (x, y) with

$$\sqrt{(x-a)^2 + (y-b)^2} = r$$

or

$$(x-a)^2 + (y-b)^2 = r^2$$

The circle with center $(0, 0)$ and radius 1, often regarded as a sort of "standardized" shape, is called the **unit circle**.

A close relative of the circle is the **ellipse**. An ellipse is defined as the set of points, the *sum* of whose distances from two **focus** points is a constant. (When the two foci are the same, the result is a circle.) If, for convenience, the focus points are taken to be $(-c, 0)$ and $(c, 0)$, and the sum of the distances is taken to be $2a$ (the factor 2 simplifies some algebra), then (x, y) is on the ellipse if and only if

$$\sqrt{(x-(-c))^2 + y^2} + \sqrt{(x-c)^2 + y^2} = 2a$$

or

$$\sqrt{(x+c)^2 + y^2} = 2a - \sqrt{(x-c)^2 + y^2}$$

or
$$x^2 + 2cx + c^2 + y^2 = 4a^2 - 4a\sqrt{(x-c)^2 + y^2} + x^2 - 2cx + c^2 + y^2$$
or
$$4(cx - a^2) = -4a\sqrt{(x-c)^2 + y^2}$$
or
$$c^2 x^2 - 2cxa^2 + a^4 = a^2(x^2 - 2cx + c^2 + y^2)$$
or
$$(c^2 - a^2)x^2 - a^2 y^2 = a^2(c^2 - a^2)$$
or
$$\frac{x^2}{a^2} + \frac{y^2}{a^2 - c^2} = 1$$

This is usually written simply
$$\frac{x^2}{a^2} + \frac{y^2}{b^2} = 1$$
where $b = \sqrt{a^2 - c^2}$ (because we must clearly choose $a > c$, it follows that $a^2 - c^2 > 0$). A picture of an ellipse is shown in the following figure.

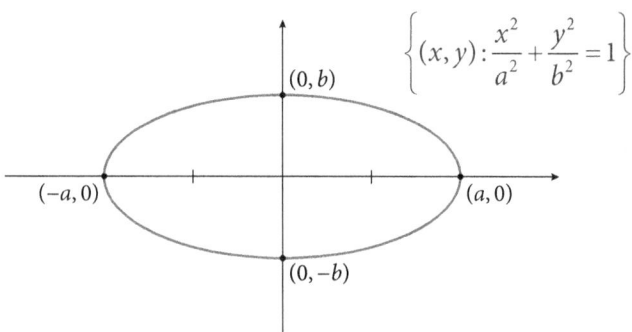

The ellipse intersects the horizontal axis when $y = 0$, so that
$$\frac{x^2}{a^2} = 1, \quad x = \pm a$$

and it intersects the vertical axis when $x = 0$, so that

$$\frac{y^2}{b^2} = 1, \quad y = \pm b$$

The **hyperbola** is defined analogously, except that we require the *difference* of the two distances to be constant. Choosing the points $(-c, 0)$ and $(c, 0)$ once again, and the constant difference as $2a$, we get, as the condition that (x, y) be on the hyperbola,

$$\sqrt{(x+c)^2 + y^2} - \sqrt{(x-c)^2 + y^2} = \pm 2a$$

which can be simplified to

$$\frac{x^2}{a^2} + \frac{y^2}{a^2 - c^2} = 1$$

In this case, however, we must clearly choose $c > a$, so that $a^2 - c^2 < 0$. If $b = \sqrt{c^2 - a^2}$, then (x, y) is on the hyperbola if and only if

$$\frac{x^2}{a^2} - \frac{y^2}{b^2} = 1$$

A picture of a hyperbola is shown in the following figure. It contains two pieces, because the difference between the distances of (x, y) from $(-c, 0)$ and $(c, 0)$ can be taken in two different orders. The hyperbola intersects the horizontal axis when $y = 0$, so that $x = \pm a$, but it never intersects the vertical axis.

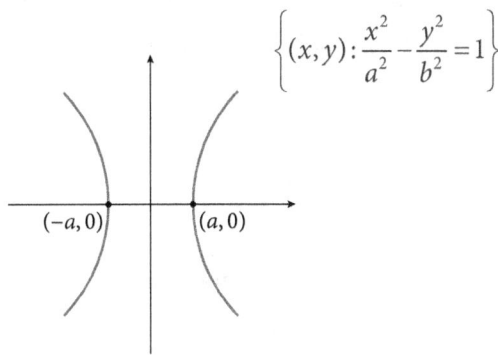

It's interesting to compare the hyperbola with $a = b = \sqrt{2}$ and the graph of the function $f(x) = 1/x$, as shown in the following figure. The drawings look quite similar, and the two sets are actually identical, except for a rotation through an angle of 45°.

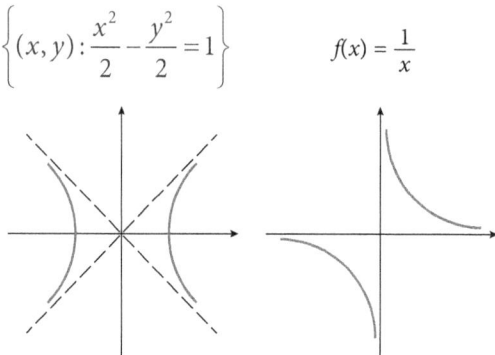

Clearly no rotation of the plane will change circles or ellipses into the graphs of functions. Nevertheless, the study of these important geometric figures can often be reduced to the study of functions. Ellipses, for example, are made up of the graphs of two functions:

$$f(x) = b\sqrt{1-(x^2/a^2)}, \quad -a \leq x \leq a$$

and

$$g(x) = -b\sqrt{1-(x^2/a^2)}, \quad -a \leq x \leq a$$

Of course, there are many other pairs of functions with this same property. For example, we can take

$$f(x) = \begin{cases} b\sqrt{1-(x^2/a^2)}, & 0 < x \leq a \\ -b\sqrt{1-(x^2/a^2)}, & -a \leq x \leq 0 \end{cases}$$

and

$$g(x) = \begin{cases} -b\sqrt{1-(x^2/a^2)}, & 0 < x \leq a \\ b\sqrt{1-(x^2/a^2)}, & -a \leq x \leq 0 \end{cases}$$

We could also choose

$$f(x) = \begin{cases} b\sqrt{1-(x^2/a^2)}, & x \text{ rational}, \quad -a \leq x \leq a \\ -b\sqrt{1-(x^2/a^2)}, & x \text{ irrational}, \quad -a \leq x \leq a \end{cases}$$

and

$$g(x) = \begin{cases} -b\sqrt{1-(x^2/a^2)}, & x \text{ rational}, \quad -a \leq x \leq a \\ b\sqrt{1-(x^2/a^2)}, & x \text{ irrational}, \quad -a \leq x \leq a \end{cases}$$

But all these other pairs necessarily involve unreasonable functions that jump around. You've probably already begun to make a distinction between those functions with reasonable graphs and those with unreasonable graphs, even if you can't state a definition of reasonable functions. Indeed, a mathematical definition of this concept is by no means easy. As you increase in mathematical sophistication, you'll be able to define more conditions that will help you recognize functions that deserve to be called reasonable.

Problems

1. Indicate on a straight line the set of all x satisfying the following conditions. Name each set by using the notation for intervals. In some cases you'll need to use the $\cup$ (union) operator.

   (a) $|x - 3| < 1$

   (b) $|x - 3| \leq 1$

   (c) $|x - a| < \varepsilon$

   (d) $|x^2 - 1| < \frac{1}{2}$

   (e) $\dfrac{1}{1+x^2} \geq \dfrac{1}{5}$

   (f) $\dfrac{1}{1+x^2} \leq a$ (Give an answer in terms of a, distinguishing various cases.)

   (g) $x^2 + 1 \geq 2$

   (h) $(x + 1)(x - 1)(x - 2) > 0$

2. This problem demonstrates a useful way to describe the points of the closed interval $[a, b]$, where $a < b$.

   (a) First consider the interval $[0, b]$, for $b > 0$. Prove that if x is in $[0, b]$, then $x = tb$ for some t with $0 \leq t \leq 1$. What's the significance of the number t? What's the midpoint of the interval $[0, b]$?

   (b) Now prove that if x is in $[a, b]$, then $x = (1 - t)a + tb$ for some t with $0 \leq t \leq 1$. (Hint: This expression can also be written as $a + t(b - a)$. What's the midpoint of the interval $[a, b]$? What's the point one-third of the way from a to b?)

3. Draw the set of all points (x, y) satisfying the following conditions. In most cases your picture will be a sizable portion of a plane, not only a line or curve.

   (a) $x > y$

   (b) $x + a > y + b$

   (c) $y < x^2$

   (d) $y \leq x^2$

   (e) $|x - y| < 1$

   (f) $|x + y| < 1$

   (g) $x + y$ is an integer

   (h) $\dfrac{1}{x + y}$ is an integer

   (i) $(x - 1)^2 + (y - 2)^2 < 1$

   (j) $x^2 < y < x^4$

4. Draw the set of all points (x, y) satisfying the following conditions.

   (a) $|x| + |y| = 1$

   (b) $|x| - |y| = 1$

   (c) $|x - 1| = |y - 1|$

(d) $|1 - x| = |y - 1|$

(e) $x^2 + y^2 = 0$

(f) $xy = 0$

(g) $x^2 - 2x + y^2 = 4$

(h) $x^2 = y^2$

5. Draw the set of all points (x, y) satisfying the following conditions. (Hint: You've already seen these answers with x and y interchanged.)

   (a) $x = y^2$

   (b) $\dfrac{y^2}{a^2} - \dfrac{x^2}{b^2} = 1$

   (c) $x = |y|$

   (d) $x = \sin y$

6. (a) Show that the straight line through (a, b) with slope m is the graph of the function $f(x) = m(x - a) + b$. This formula, known as the **point-slope form**, is far more convenient than the equivalent expression $f(x) = mx + (b - ma)$; it's immediately clear from the point-slope form that the slope is m, and that the value of f at a is b.

   (b) For $a \neq c$, show that the straight line through (a, b) and (c, d) is the graph of the function
   $$f(x) = \frac{d-b}{c-a}(x-a) + b$$

   (c) When are the graphs of $f(x) = mx + b$ and $g(x) = m'x + b'$ parallel straight lines?

7. (a) For any numbers A, B, and C, with A and B not both 0, show that the set of all (x, y) satisfying $Ax + By + C = 0$ is a straight line (possibly a vertical one). (Hint: First decide when a vertical straight line is described.)

   (b) Show conversely that every straight line, including vertical ones, can be described as the set of all (x, y) satisfying $Ax + By + C = 0$.

8. (a) Prove that the graphs of the functions

$$f(x) = mx + b$$
$$g(x) = nx + c$$

are perpendicular if $mn = -1$, by computing the squares of the lengths of the sides of the triangle in the following figure. (Why is this special case, where the lines intersect at the origin, as good as the general case?)

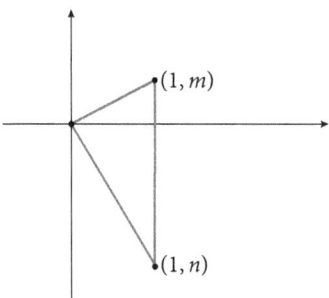

(b) Prove that the two straight lines consisting of all (x, y) satisfying the conditions

$$Ax + By + C = 0$$
$$A'x + B'y + C' = 0$$

are perpendicular if and only if $AA' + BB' = 0$.

9. Sketch the graphs of the following functions, plotting enough points to get a good idea of the general appearance. Part of the problem is to make a reasonable decision how many is "enough"; the examples posed below are meant to show that a little thought will often be more valuable than hundreds of individual points.

(a) $f(x) = x + \dfrac{1}{x}$

What happens for x near 0, and for large x? Where does the graph lie in relation to the graph of the identity function? Why does it suffice to consider only positive x at first?

(b) $f(x) = x - \dfrac{1}{x}$

(c) $f(x) = x^2 + \dfrac{1}{x^2}$

(d) $f(x) = x^2 - \dfrac{1}{x^2}$

10. Describe the general features of the graph of f if

   (a) f is even (that is, $f(x) = f(-x)$ for all x)

   (b) f is odd (that is, $-f(x) = f(-x)$ for all x)

   (c) f is nonnegative

   (d) $f(x) = f(x + a)$ for all x (a function with this property is called **periodic**, with **period** a)

11. Graph the functions $f(x) = \sqrt[m]{x}$ for $m = 1, 2, 3, 4$.

   There's an easy way to do this by using the figure for power functions. Keep in mind that $\sqrt[m]{x}$ means the *positive* mth root of x when m is even. Also note that there will be an important difference between the graphs when m is even and when m is odd.

12. (a) Graph $f(x) = |x|$ and $f(x) = x^2$.

    (b) Graph $f(x) = |\sin x|$ and $f(x) = \sin^2 x$.

    What's the important difference between the first graph and the second graph in both parts (a) and (b)?

13. Describe the graph of g in terms of the graph of f if

    (a) $g(x) = f(x) + c$

    (b) $g(x) = f(x + c)$ (be careful)

    (c) $g(x) = cf(x)$ (distinguish the cases $c = 0$, $c > 0$, $c < 0$)

    (d) $g(x) = f(cx)$ (distinguish the cases $c = 0$, $c > 0$, $c < 0$)

    (e) $g(x) = f(1/x)$

    (f) $g(x) = f(|x|)$

    (g) $g(x) = |f(x)|$

(h) $g(x) = \max(f, 0)$

(i) $g(x) = \min(f, 0)$

(j) $g(x) = \max(f, 1)$

14. Draw the graph of $f(x) = ax^2 + bx + c$.

15. Suppose that A and C are not both 0. Show that the set of all (x, y) satisfying
$$Ax^2 + Bx + Cy^2 + Dy + E = 0$$
is either a parabola, an ellipse, or a hyperbola (or possibly 0). (Hint: The case $C = 0$ is essentially the preceding problem, and the case $A = 0$ is a minor variant of that. Also consider separately the cases where A and B are both positive, are both negative, and have opposite signs.)

16. The symbol $[x]$ denotes the largest integer that is $\leq x$. Thus, $[2.1] = [2] = 2$ and $[-0.9] = [-1] = -1$. Draw the graph of the following functions.

    (a) $f(x) = [x]$
    (b) $f(x) = x - [x]$
    (c) $f(x) = \sqrt{x - [x]}$
    (d) $f(x) = [x] + \sqrt{x - [x]}$

17. Draw the graph of $f(x) = \{x\}$, where $\{x\}$ is defined to be the distance from x to the nearest integer.

18. Use your intuitive notion of infinite decimals to describe as best you can the graphs of the following functions.

    (a) $f(x)$ = the 1st number in the decimal expansion of x.
    (b) $f(x)$ = the 2nd number in the decimal expansion of x.

19. (a) The points on the graph of $f(x) = x^2$ are the ones of the form (x, x^2). Prove that each such point is equidistant from the point $(0, ¼)$ and the graph of $g(x) = -¼$, as shown in the following figure.

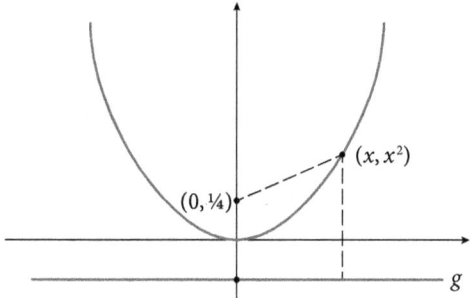

(b) Given a point $P = (\alpha, \beta)$ and a horizontal line L, the graph of $g(x) = y$, show that the set of all points (x, y) equidistant from P and L is the graph of a function of the form $f(x) = ax^2 + bx + c$.

5 Solutions

Chapter 1

1. (a) $\{\ldots, -11, -6, -1, 4, 9, 14, 19, \ldots\}$
   (b) $\{-2, -1, 0, 1, 2, 3, 4, 5, 6\}$
   (c) $\{-\sqrt{3}, \sqrt{3}\}$
   (d) $\{-2, -3\}$
   (e) $\{\ldots, -3, -2, -1, 0, 1, 2, 3, \ldots\} = \mathbb{Z}$
   (f) $\{-4, -3, -2, -1, 0, 1, 2, 3, 4\}$
   (g) $\{0\}$
   (h) $\{\ldots, -3, -2, -1, 0, 1, 2, 3, \ldots\} = \mathbb{Z}$

2. (a) $\{2^x : x \in \mathbb{N}\}$
   (b) $\{3x : x \in \mathbb{Z}\}$
   (c) $\{x^2 : x \in \mathbb{Z}\}$
   (d) $\{x \in \mathbb{Z} : 3 \le x \le 8\} = \{x \in \mathbb{N} : 3 \le x \le 8\}$
   (e) $\{2^n : n \in \mathbb{Z}\}$
   (f) $\{k\pi/2 : k \in \mathbb{Z}\}$

3. (a) 3
   (b) 1
   (c) 19
   (d) 7
   (e) 0

4. (a)

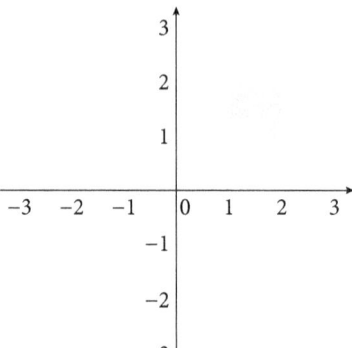

(b)

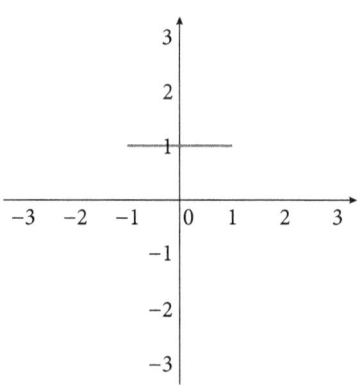

(c)

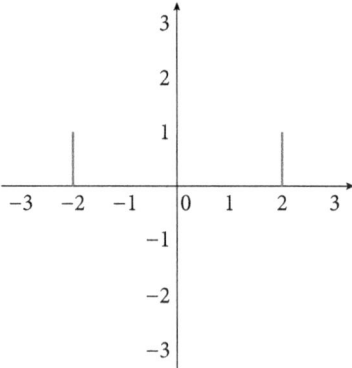

(d)
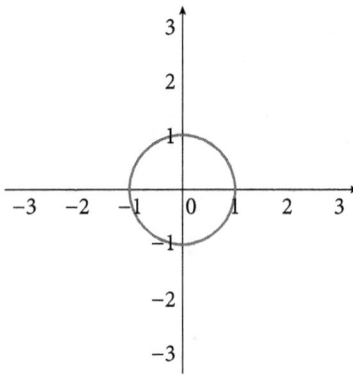

(e)
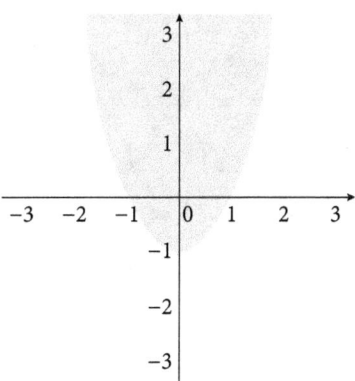

(f)
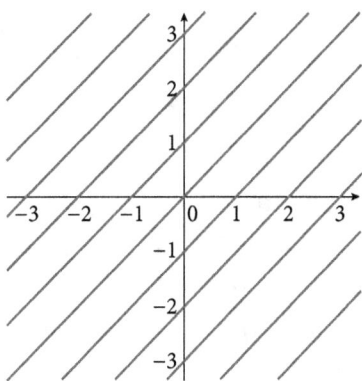

(g)

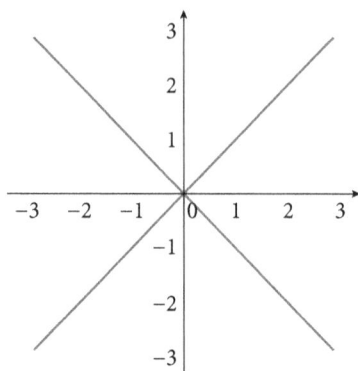

5. (a) $A \times B = \{(1, a), (1, c), (2, a), (2, c), (3, a), (3, c), (4, a), (4, c)\}$

   (b) $B \times A = \{(a, 1), (a, 2), (a, 3), (a, 4), (c, 1), (c, 2), (c, 3), (c, 4)\}$

   (c) $A \times A =$
   {
   (1, 1), (1, 2), (1, 3), (1, 4),
   (2, 1), (2, 2), (2, 3), (2, 4),
   (3, 1), (3, 2), (3, 3), (3, 4),
   (4, 1), (4, 2), (4, 3), (4, 4)
   }

   (d) $B \times B = \{(a, a), (a, c), (c, a), (c, c)\}$

   (e) $\emptyset \times B = \{(a, b) : a \in \emptyset, b \in B\} = \emptyset$ (There are no ordered pairs (a, b) with $a \in \emptyset$)

   (f) $(A \times B) \times B =$
   {
   ((1, a), a), ((1, c), a), ((2, a), a), ((2, c), a),
   ((3, a), a), ((3, c), a), ((4, a), a), ((4, c), a),
   ((1, a), c), ((1, c), c), ((2, a), c), ((2, c), c),
   ((3, a), c), ((3, c), c), ((4, a), c), ((4, c), c)
   }

(g) $A \times (B \times B) =$
{
$(1, (a, a)), (1, (a, c)), (1, (c, a)), (1, (c, c)),$
$(2, (a, a)), (2, (a, c)), (2, (c, a)), (2, (c, c)),$
$(3, (a, a)), (3, (a, c)), (3, (c, a)), (3, (c, c)),$
$(4, (a, a)), (4, (a, c)), (4, (c, a)), (4, (c, c))$
}

(h) $B^3 = \{(a, a, a), (a, a, c), (a, c, a), (a, c, c), (c, a, a), (c, a, c), (c, c, a), (c, c, c)\}$

6. (a) $\{(-\sqrt{2}, a), (\sqrt{2}, a), (-\sqrt{2}, c), (\sqrt{2}, c), (-\sqrt{2}, e), (\sqrt{2}, e)\}$
   (b) $\{(-\sqrt{2}, -2), (\sqrt{2}, 2), (-\sqrt{2}, 2), (\sqrt{2}, -2)\}$
   (c) $\{(\emptyset, 0, 0), (\emptyset, 0, 1), (\emptyset, \emptyset, 0), (\emptyset, \emptyset, 1)\}$

7. (a)

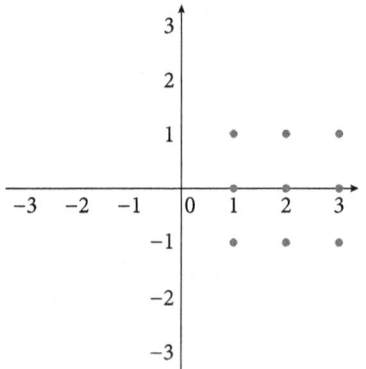

(b)

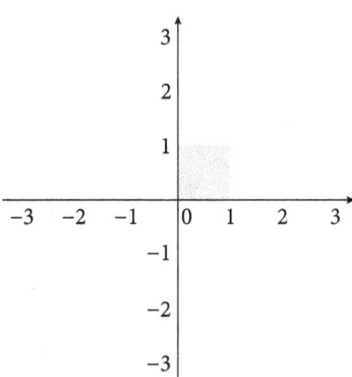

(c)

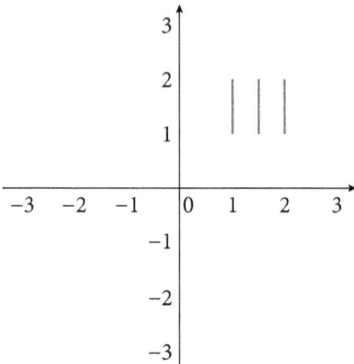

(d)

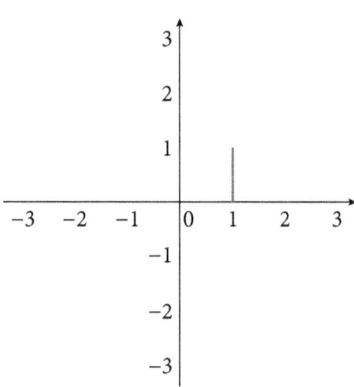

(e)

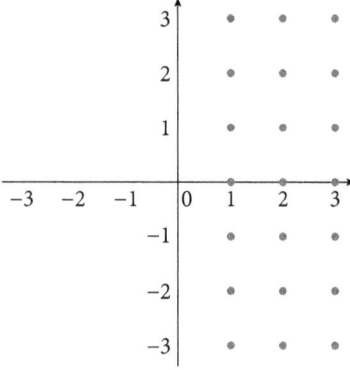

(f)

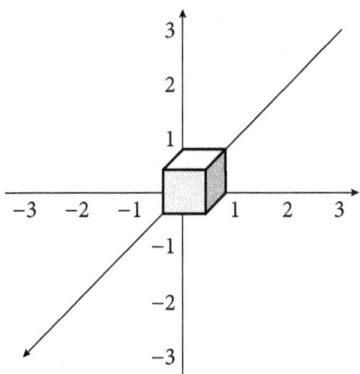

8. (a) The subsets of {1, 2, 3, 4} are {}, {1}, {2}, {3}, {4}, {1, 2}, {1, 3}, {1, 4}, {2, 3}, {2, 4}, {3, 4}, {1, 2, 3}, {1, 2, 4}, {1, 3, 4}, {2, 3, 4}, and {1, 2, 3, 4}.

   (b) The subsets of {{ℝ}} are {} and {{ℝ}}.

   (c) The subsets of {∅} are {} and {∅}.

   (d) The subsets of {ℝ, {ℚ, ℕ}} are {}, {ℝ}, {{ℚ, ℕ}}, and {ℝ, {ℚ, ℕ}}.

9. (a) {{3, 2}, {3, a}, {2, a}}

   (b) ∅

10. (a) True. Every set is a subset of itself.

    (b) False. The elements of ℝ² (ordered pairs) don't come from ℝ³ (ordered triples).

    (c) True.

    (d) False.

11. (a) {∅, {{a, b}}, {{c}}, {{a, b}, {c}}}

    (b) {∅, {{∅}}, {5}, {{∅}, 5}}

    (c) {∅, {∅}, {{2}}, {∅, {2}}}

(d) {(∅, ∅), (∅, {0}), (∅, {1}), (∅, {0, 1}),
({a}, ∅), ({a}, {0}), ({a}, {1}), ({a}, {0,1}),
({b}, ∅), ({b}, {0}), ({b}, {1}), ({b}, {0,1}),
({a, b}, ∅), ({a, b}, {0}), ({a, b}, {1}), ({a, b}, {0, 1})}

(e) {∅, {(a, 0)}, {(b, 0)}, {(a, 0), (b, 0)}}

(f) {∅, {∅}, {{1}}, {{2}}, {{3}}, {{1, 2}}, {{1, 3}}, {{2, 3}}, {{1, 2, 3}}}

12. (a) $2^{2^{2^m}}$
    (b) 2^{mn}
    (c) $m + 1$
    (d) $|P(P(P(A \times \emptyset)))| = |P(P(P(\emptyset)))| = 4$

13. (a) {1, 3, 4, 5, 6, 7, 8, 9}
    (b) {4, 6}
    (c) {3, 7, 1, 9}
    (d) {3, 6, 7, 1, 9}
    (e) {5, 8}
    (f) {4}
    (g) {5, 8, 4}
    (h) {5, 6, 8, 4}
    (i) ∅

14. (a) {(1, 1), (1, 2)}
    (b) {(0, 1), (0, 2), (1, 1), (1, 2), (2, 1), (2, 2)}
    (c) {(0, 1), (0, 2)}
    (d) {(1, 0), (1, 1)}
    (e) ∅
    (f) {∅, {1}}
    (g) {{0}, {0, 1}}
    (h) {{}, {1}}
    (i) {∅, {(0, 1)}, {(0, 2)}, {(1, 1)}, {(1, 2)}, {(0, 1), (0, 2)}, {(0, 1), (1, 1)}, {(0, 1), (1, 2)}, {(0, 2), (1, 1)}, {(0, 2), (1, 2)}, {(1, 1), (1, 2)}, {(0, 2), (1, 1), (1, 2)}, {(0, 1), (1, 1), (1, 2)}, {(0, 1), (0, 2), (1, 2)}, {(0, 1), (0, 2), (1, 1)}, {(0, 1), (0, 2), (1, 1), (1, 2)}}

15.

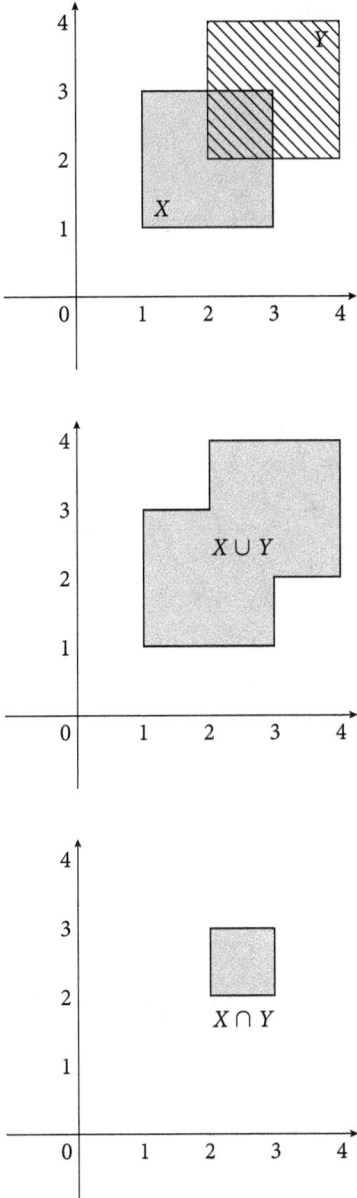

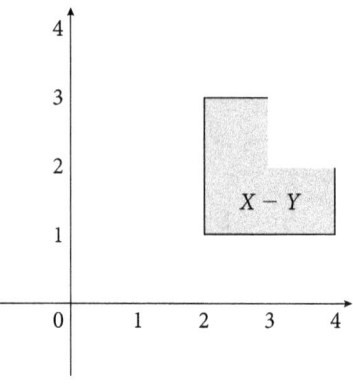

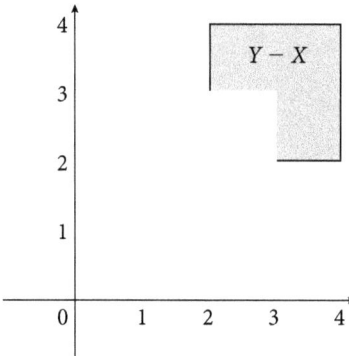

16.

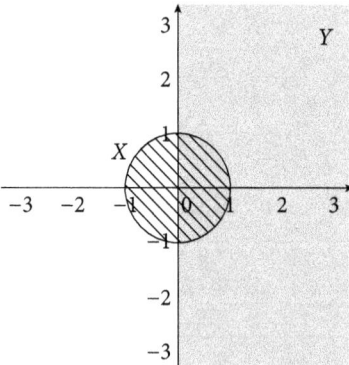

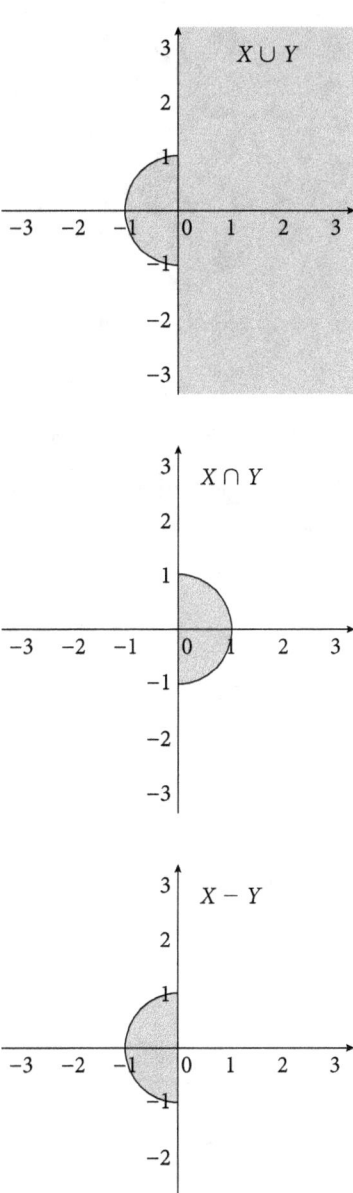

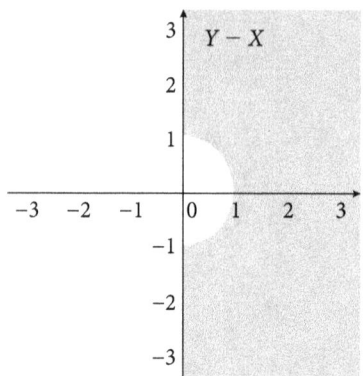

17. (a) True. Draw a picture to convince yourself if necessary.

    (b) False. Notice for example that (0.5, 0.5) is in the right-hand set, but not in the left-hand set.

18. (a) {0, 2, 5, 8, 10}
    (b) {0, 1, 2, 3, 7, 9, 10}
    (c) ∅
    (d) {0, 1, 2, 3, 4, 5, 6, 7, 8, 9, 10} = U
    (e) A
    (f) {4, 6}
    (g) {5, 8}
    (h) {5, 8}
    (i) {0, 1, 2, 3, 4, 6, 7, 9, 10}

19.

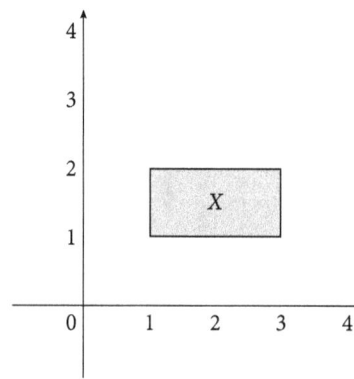

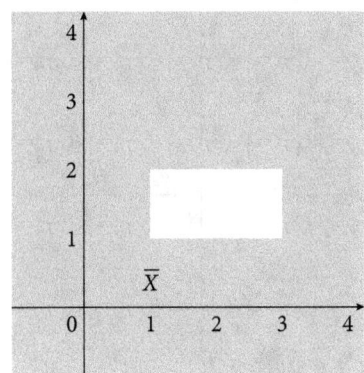

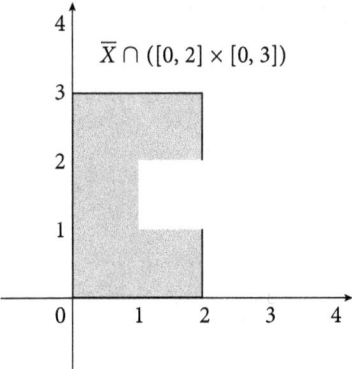

20.

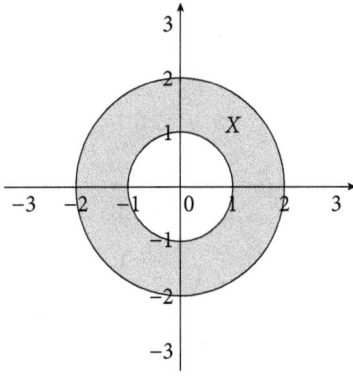

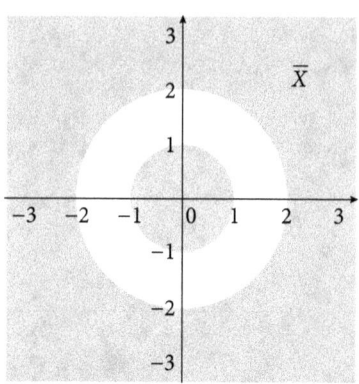

21.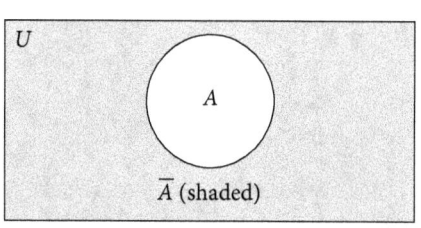

22.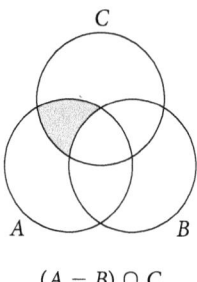

$(A - B) \cap C$

23. The Venn diagrams are the same for both expressions, so we're inclined to say that $A \cup (B \cap C) = (A \cup B) \cap (A \cup C)$ holds for all sets A, B, and C.

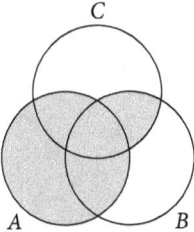

$A \cup (B \cap C) = (A \cup B) \cap (A \cup C)$

24. The Venn diagrams are the same for both expressions, so we're inclined to say that $\overline{A \cap B} = \overline{A} \cup \overline{B}$ for all sets A and B.

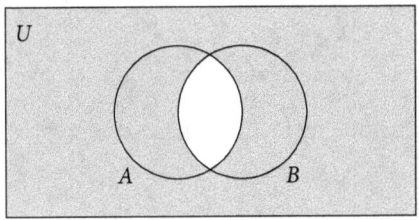

$\overline{A \cap B} = \overline{A} \cup \overline{B}$

25.

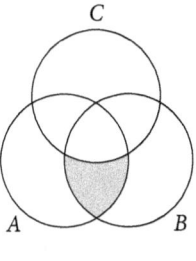

$(A \cap B) - C$

26. The simplest answer is $(B \cap C) - A$.

27. One answer is $(A \cup B \cup C) - (A \cap B \cap C)$.

28. (a) $\{a, b, c, d, e, f, g, h\}$
    (b) $\{a, b\}$

29. (a) $\{0\} \cup \mathbb{N}$
    (b) $\{0, 1\}$

30. (a) $[1, \infty)$
    (b) $\emptyset$

31. (a) $\{(x, y) : x, y \in \mathbb{R}, y \geq 1\}$
    (b) $\emptyset$

32. (a) $\mathbb{N}$
    (b) $\emptyset$

33. Yes, this expression is always true.

    The first expression is true and the second expression is false.

Chapter 2
1. (a) $1 = a^{-1}a = a^{-1}(ax) = (a^{-1}a)x = 1 \cdot x = x$.

   (b) $(x - y)(x + y) = [x + (-y)](x + y)$
   $= x(x + y) + (-y)(x + y)$
   $= x(x + y) - [y(x + y)]$
   $= x^2 + xy - [yx + y^2]$
   $= x^2 + xy - yx - y^2$
   $= x^2 - y^2$.

   (c) If $x^2 = y^2$, then $0 = x^2 - y^2 = (x - y)(x + y)$, so either $x - y = 0$ or $x + y = 0$; that is, either $x = -y$ or $x = y$.

   (d) $(x - y)(x^2 + xy + y^2) = x(x^2 + xy + y^2) - [y(x^2 + xy + y^2)]$
   $= x^3 + x^2y + xy^2 - [yx^2 + xy^2 + y^3]$
   $= x^3 - y^3$.

   (e) $(x - y)(x^{n-1} + x^{n-2}y + \cdots + xy^{n-2} + y^{n-1})$
   $= x(x^{n-1} + x^{n-2}y + \cdots + xy^{n-2} + y^{n-1}) - [y(x^{n-1} + x^{n-2}y + \cdots + xy^{n-2} + y^{n-1})]$
   $= x^n + x^{n-1}y + \cdots + x^2y^{n-2} + xy^{n-1} - [x^{n-1}y + x^{n-2}y^2 + \cdots + xy^{n-1} + y^n]$
   $= x^n - y^n$.

   By using summation notation, this proof can be written as

   $(x - y) \cdot \left(\sum_{j=0}^{n-1} x^j y^{n-1-j} \right) = x \left(\sum_{j=0}^{n-1} x^j y^{n-1-j} \right) - \left[y \left(\sum_{j=0}^{n-1} x^j y^{n-1-j} \right) \right]$

   $= x^n + \sum_{j=0}^{n-2} x^{j+1} y^{n-1-j} - \left[\sum_{j=1}^{n-1} x^j y^{n-j} + y^n \right]$

   $= x^n + \sum_{j=0}^{n-2} x^{j+1} y^{n-1-j} - \left[\sum_{k=0}^{n-2} x^{k+1} y^{n-(k+1)} + y^n \right]$

   $= x^n - y^n$ (letting $k = j - 1$)

   (f) Replace y by $-y$ in part (d).

2. One step requires dividing by $x - y = 0$.

3. (a) $ab(a^{-1}b^{-1}) = (a \cdot a^{-1})(b \cdot b^{-1}) = 1$, so $a^{-1} \cdot b^{-1} = (ab)^{-1}$.

   (b) $a/b = ab^{-1} = (ac)(b^{-1}c^{-1}) = (ac)(bc)^{-1}$ [by part (a)] $= ac/bc$.

   (c) $(ad + bc)/(bd) = (ad + bc)(bd)^{-1} = (ad + bc)(b^{-1}d^{-1})$ [by part (a)] $= ab^{-1} + cd^{-1} = a/b + c/d$.

   (d) $(a/b)(c/d) = (ab^{-1})(cd^{-1}) = (ac)(b^{-1}d^{-1}) = (ac)(bd)^{-1}$ [by part (a)] $= (ac)/(bd)$.

   (e) $(a/b)/(c/d) = (a/b)(c/d)^{-1} = (a \cdot b^{-1})(c \cdot d^{-1})^{-1} = (a \cdot b^{-1})(c^{-1} \cdot d) = ad(b^{-1} \cdot c^{-1}) = ad(bc)^{-1} = (ad)/(bc)$.

   (f) If $ab^{-1} = cd^{-1}$, then $(ab^{-1})bd = (cd^{-1})bd$, or $ad = bc$. Conversely, if $ad = bc$, then $(ad)d^{-1}b^{-1} = (bc)d^{-1}b^{-1}$, or $ab^{-1} = cd^{-1}$. If $ab^{-1} = ba^{-1}$, then $a^2 = b^2$, so by Problem 1(c), $a = b$ or $a = -b$. Conversely, if $a = b$, then $a/b = b/a = 1$ and if $a = -b$, then $a/b = b/a = -1$.

4. (a) $x < -1$.

   (b) All x.

   (c) $x > \sqrt{7}$ or $x < -\sqrt{7}$.

   (d) $x > 3$ or $x < 1$.

   (e) All x, because $x^2 - 2x + 2 = (x - 1)^2 + 1$.

   (f) $x > [-1+\sqrt{5}]/2$ or $x < [-1-\sqrt{5}]/2$.

   (g) $x > 3$ or $x < -2$, because 3 and -2 are the roots of $x^2 - x - 6 = 0$.

   (h) All x, because $x^2 + x + 1 = [x + (½)]^2 + ¾$.

   (i) $x > \pi$ or $-5 < x < 3$.

   (j) $x > \sqrt{2}$ or $x < \sqrt[3]{2}$.

   (k) $x < 3$.

   (l) $x < 1$.

   (m) $x > 1$ or $0 < x < 1$.

   (n) $x > 1$ or $x < -1$.

5. (a) $b - a$ and $d - c$ are in P, so $(b - a) + (d - c) = (b + d) - (a + c)$ is in P. Thus, $b + d > a + c$.

    (b) $b - a$ is in P, so $-a - (-b)$ is in P.

    (c) Using part (b), $-c < -d$; then part (a) implies that $a + (-c) < b + (-d)$.

    (d) $b - a$ is in P and c is in P, so $c(b - a) = bc - ac$ is in P.

    (e) $(b - a)$ and $-c$ are in P, so $-c(b - a) = ac - bc$ is in P; that is, $ac > bc$.

    (f) If $a > 1$, then $a > 0$, so $a^2 > a \cdot 1$, by part (d).

    (g) Using part (d), $a > 0$ and $a < 1$, so $a^2 < a$.

    (h) If $a = 0$ or $c = 0$, then $ac = 0$, but $bd > 0$, so $ac < bd$. Otherwise, we have $ac < bc < bd$ by applying part (d) twice.

    (i) Substitute a for c and b for d in part (h).

    (j) If $a < b$ were false, then either $a = b$ or $a > b$. But if $a = b$, then $a^2 = b^2$, and if $a > b \geq 0$, then $a^2 > b^2$, by part (i).

6. (a) From $0 \leq x < y$ and Problem 5(h) we have $x^2 < y^2$ [as in Problem 5(i)]. Then from $0 \leq x < y$ and $x^2 < y^2$ we have $x^3 < y^3$. We can continue in this way to prove that $x^n < y^n$ for $n = 2, 3, \ldots$. (A rigorous proof uses induction.)

    (b) If $0 \leq x < y$, then $x^n < y^n$ by part (a). If $x < y \leq 0$, then $0 \leq -y < -x$, so $(-y)^n < (-x)^n$ by part (a); this result means that $-y^n < -x^n$ (because n is odd) and hence $x^n < y^n$. Finally, if $x < 0 \leq y$, then $x^n < 0 \leq y^n$ (because n is odd). Thus, in all cases, if $x < y$, then $x^n < y^n$.

    (c) This follows immediately from part (b), because $x < y$ would imply that $x^n < y^n$, while $y < x$ would imply that $y^n < x^n$.

    (d) Similarly, if n is even, then by using part (a) instead of part (b) we see that if $x, y \geq 0$ and $x^n = y^n$, then $x = y$. Moreover, if $x, y \leq 0$ and $x^n = y^n$, then $-x, -y \geq 0$ and $(-x)^n = (-y)^n$, so again $x = y$. The only other possibility is that one of x and y is positive, the other negative. In this case x and $-y$ are both positive or both negative. Moreover

$x^n = (-y)^n$, because n is even, so it follows from the previous cases that $x = -y$.

7. (a) $\left|\sqrt{2}+\sqrt{3}-\sqrt{5}+\sqrt{7}\right|$.

   (b) $|a| + |b| - |a + b|$.

   (c) $|a + b| + |c| - |a + b + c|$.

   (d) $x^2 - 2xy + y^2$.

   (e) $\left|\left(\left|\sqrt{2}+\sqrt{3}\right|-\left|\sqrt{5}-\sqrt{7}\right|\right)\right|$.

8. (a) a if $a \geq -b$ and $b \geq 0$;
   $-a$ if $a \leq -b$ and $b \leq 0$;
   $a + 2b$ if $a \geq -b$ and $b \leq 0$;
   $-a - 2b$ if $a \leq -b$ and $b \geq 0$.

   (b) $x - 1$ if $x \geq 1$;
   $1 - x$ if $0 \leq x \leq 1$;
   $1 + x$ if $-1 \leq x \leq 0$;
   $-1 - x$ if $x \leq -1$.

   (c) $x - x^2$ if $x \geq 0$;
   $-x - x^2$ if $x \leq 0$.

   (d) a if $a \geq 0$;
   $3a$ if $a \leq 0$.

9. (a) $x = 11, -5$.

   (b) $-5 < x < 11$.

   (c) $-6 < x < -2$.

   (d) $x < 1$ or $x > 2$ (the distance from x to 1 plus the distance from x to 2 equals 1 precisely when $1 \leq x \leq 2$).

   (e) No x (the distance from x to 1 plus the distance from x to -1 is at least 2).

   (f) No x.

   (g) $x = 1, -1$.

(h) If $x > 1$ or $x < -2$, then the condition becomes $(x - 1)(x + 2) = 3$, or $x^2 + x - 5 = 0$, for which the solutions are $(-1 + \sqrt{21})/2$ and $(-1 - \sqrt{21})/2$. Because the first solution is > 1 and the second is < -2, both are solutions to the equation $|x - 1| \cdot |x + 2| = 3$. For $-2 < x < 1$ the condition becomes $(1 - x)(x + 2) = 3$ or $x^2 + x + 1 = 0$, which has no solutions.

10. (a) $(|xy|)^2 = (xy)^2 = x^2y^2 = |x|^2|y|^2 = (|x| \cdot |y|)^2$; because $|xy|$ and $|x| \cdot |y|$ are both ≥ 0, this proves that $|xy| = |x| \cdot |y|$.

(b) $|1/x| \cdot |x| = |(1/x) \cdot x|$ [by part (a)] $= |1| = 1$, so $|1/x| = 1/|x|$.

(c) $|x|/|y| = |x| \cdot |y|^{-1} = |x| \cdot |y^{-1}|$ [by part (b)] $= |xy^{-1}|$ [by part (a)] $= |x/y|$.

(d) $|x - y| = |x + (-y)| \leq |x| + |-y| = |x| + |y|$.

(e) It follows from part (d) that $|x| = |y - (y - x)| \leq |y| + |y - x|$, so $|x| - |y| \leq |x - y|$.

(f) Interchanging x and y in part (e) gives $|y| - |x| \leq |x - y|$. Combining this with part (e) yields $|(|x| - |y|)| \leq |x - y|$.

(g) $|x + y + z| \leq |x + y| + |z| \leq |x| + |y| + |z|$. If equality holds, then $|x + y| = |x| + |y|$, so x and y have the same sign. Moreover, z must have the same sign as $x + y$, so x, y, and z must all have the same sign (unless one is 0).

11. If $a \geq 0$, then $|a| = a = -(-a) = |-a|$, because $-a \leq 0$. The equality is proved for $a \leq 0$ by replacing a by $-a$.

12. If $|a| \leq b$, then clearly $b \geq 0$. Now $|a| \leq b$ means that $a \leq b$ if $a \geq 0$, and surely $a \leq b$ if $a \leq 0$. Similarly, $|a| \leq b$ means $-a \leq b$, and hence $-b \leq a$, if $a \leq 0$, and surely $-b \leq a$ if $a \geq 0$. So $-b \leq a \leq b$.

Conversely, if $-b \leq a \leq b$, then $|a| = a \leq b$ if $a \geq 0$, while $|a| = -a \leq b$ if $a \leq 0$.

13. From $-|a| \leq a \leq |a|$ and $-|b| \leq b \leq |b|$ it follows that

$$-(|a| + |b|) \leq a + b \leq |a| + |b|$$

so $|a + b| \leq |a| + |b|$.

14. Because
$$2x^2 - 3x + 4 = 2\left(x - \frac{3}{4}\right)^2 + 4 - \frac{9}{8}$$
$$= 2\left(x - \frac{3}{4}\right)^2 + \frac{23}{8}$$
the smallest possible value is $23/8$, when $(x - 3/4)^2 = 0$, or $x = 3/4$.

15. We have
$$x^2 - 3x + 2y^2 + 4y + 2 = \left(x - \frac{3}{2}\right)^2 + 2(y+1)^2 - \frac{9}{4}$$
so the smallest possible value is $-9/4$, when $x = 3/2$ and $y = -1$.

16. For each y we have
$$x^2 + 4xy + 5y^2 - 4x - 6y + 7 = x^2 + 4(y-1)x + 5y^2 - 6y + 7$$
$$= [x + 2(y-1)]^2 + 5y^2 - 6y + 7 - 4(y-1)^2$$
$$= [x + 2(y-1)]^2 + (y+1)^2 + 2$$
so the smallest possible value is 2, when $y = -1$ and $x = -2(y - 1) = 4$.

17. (a) is a straightforward check.

   (b) We have
$$x^2 + bx + c = \left(x + \frac{b}{2}\right)^2 + \left(c - \frac{b^2}{4}\right) \geq c - \frac{b^2}{4}$$
   but $c - b^2/4 > 0$, so $x^2 + bx + c > 0$ for all x.

   (c) Apply part (b) with y for b and y^2 for c: we have $b^2 - 4c = y^2 - 4y^2 < 0$ for $y \neq 0$, so $x^2 + xy + y^2 > 0$ for all x, if $y \neq 0$ (and surely $x^2 + xy + y^2 > 0$ for all $x \neq 0$ if $y = 0$).

   (d) α must satisfy $(\alpha y)^2 - 4y^2 < 0$, or $\alpha^2 < 4$, or $|\alpha| < 2$.

18. P2, P3, P4, P6, P7, and P8 are obvious from a glance at the tables. There are eight cases for P1, and even this number can be reduced: because P2 is true, it's clear that $a + (b + c) = (a + b) + c$ if a, b, or c is 0, so only the case $a = b = c = 1$ must be checked. Similarly for P5.

Finally, P9 is true for $a = 0$, because $0 \cdot b = 0$ for all b, and for $a = 1$, because $1 \cdot b = b$ for all b.

19. Because $1^2 = 1 \cdot (2) \cdot (2 \cdot 1 + 1)/6$, the formula is true for $n = 1$. Suppose that the formula is true for k. Then

$$\begin{aligned}1^2 + \cdots + k^2 + (k+1)^2 &= \frac{k(k+1)(2k+1)}{6} + (k+1)^2 \\ &= \frac{(k+1)}{6} + [k(2k+1) + 6(k+1)] \\ &= \frac{(k+1)}{6} + [(k+2)(2k+3)] \\ &= \frac{(k+1)(k+2)(2[k+1]+1)}{6}\end{aligned}$$

so the formula is true for $k + 1$.

20. Because $1^3 = 1^2$, the formula is true for $n = 1$. Suppose that the formula is true for k. Then

$$\begin{aligned}(1 + \cdots + k + [k+1])^2 &= (1 + \cdots + k)^2 + 2(1 + \cdots + k)(k+1) + (k+1)^2 \\ &= 1^3 + \cdots + k^3 + 2\frac{k(k+1)}{2}(k+1) + (k+1)^2 \\ &= 1^3 + \cdots + k^3 + (k^3 + 2k^2 + k) + (k^2 + 2k + 1) \\ &= 1^3 + \cdots + k^3 + (k+1)^3\end{aligned}$$

so the formula is true for $k + 1$.

21. $\displaystyle\sum_{i=1}^{n}(2i-1) = 1 + 3 + 5 + \cdots + (2n-1)$

$$\begin{aligned}&= 1 + 2 + 3 + \cdots + 2n - 2(1 + \cdots + n) \\ &= \frac{(2n)(2n+1)}{2} - n(n+1) \\ &= n^2\end{aligned}$$

22. $\sum_{i=1}^{n}(2i-1)^2 = 1^2 + 3^2 + \cdots + (2n-1)^2$

$= [1^2 + 2^2 + \cdots + (2n)^2] - [2^2 + 4^2 + 6^2 + \cdots + (2n)^2]$
$= [1^2 + 2^2 + \cdots + (2n)^2] - 4[1^2 + 2^2 + 3^2 + \cdots + (n)^2]$
$= \dfrac{2n(2n+1)(4n+1)}{6} - \dfrac{4n(n+1)(2n+1)]}{6}$
$= \dfrac{2n(2n+1)[4n+1-2(n+1)]}{6}$
$= \dfrac{n(2n+1)(2n-1)}{3}$

23. $\binom{n}{k-1} + \binom{n}{k} = \dfrac{n!}{(k-1)!(n-k+1)!} + \dfrac{n!}{k!(n-k)!}$

$= \dfrac{k(n!)}{k!(n+1-k)!} + \dfrac{(n+1-k)n!}{k!(n+1-k)!}$
$= \dfrac{(n+1)n!}{k!(n+1-k)!}$
$= \binom{n+1}{k}$

24. Because

$$1+r = \dfrac{1-r^2}{1-r}$$

the formula is true for $n = 1$. Suppose that

$$1+r+\cdots+r^n = \dfrac{1-r^{n+1}}{1-r}$$

134 Essential Advanced Precalculus

Then
$$1+r+\cdots+r^n+r^{n+1} = \frac{1-r^{n+1}}{1-r}+r^{n+1}$$
$$= \frac{1-r^{n+1}+r^{n+1}(1-r)}{1-r}$$
$$= \frac{1-r^{n+2}}{1-r}$$

25.
$$S = 1+r+\cdots+r^n$$
$$rS = r+\cdots+r^n+r^{n+1}$$

Thus
$$S - rS = S(1-r) = 1 - rn + 1$$

so
$$S = \frac{1-r^{n+1}}{1-r}$$

26. 1 is either even or odd, in fact it's odd. Suppose that n is either even or odd; then n can be written either as $2k$ or $2k + 1$. In the first case $n + 1 = 2k + 1$ is odd; in the second case $n + 1 = 2k + 1 + 1 = 2(k + 1)$ is even. In either case, $n + 1$ is either even or odd. (This proof looks sketchy, but it's correct.)

27. Yes, for if $a + b$ were rational, then $b = (a + b) - a$ would be rational. If a and b are irrational, then $a + b$ could be rational, for b could be $r - a$ for some rational number a.

28. If $a = 0$, then ab is rational. But if $a \neq 0$, then ab can't be rational, for then $b = (ab) \cdot a^{-1}$ would be rational.

29. Yes; for example, $\sqrt[4]{2}$.

30. Yes; for example, $\sqrt{2}$ and $-\sqrt{2}$.

31. Because $a^{n+1} = a^n \cdot a = a^n \cdot a^1$, the first equation is true for $m = 1$. Suppose that $a^{n+m} = a^n \cdot a^m$. Then

$$\begin{aligned} a^{n+(m+1)} = a^{(n+m)+1} &= a^{n+m} \cdot a \quad \text{(by definition)} \\ &= (a^n \cdot a^m) \cdot a \\ &= a^n \cdot (a^m \cdot a) \\ &= a^n \cdot a^{m+1} \quad \text{(by definition)} \end{aligned}$$

so the first equation is true for $m + 1$.

Because $(a^n)^1 = a^n = a^{n \cdot 1}$, the second equation is true for $m = 1$. Suppose that $(a^n)^m = a^{nm}$. Then

$$\begin{aligned} (a^n)^{m+1} &= (a^n)^m \cdot a^n \quad \text{(by definition)} \\ &= a^{nm} \cdot a^n \\ &= a^{nm+n} \quad \text{(by above)} \\ &= a^{n(m+1)} \end{aligned}$$

Chapter 3

1. (a) $(x+1)/(x+2)$; the expression $f(f(x))$ makes sense only when $x \neq -1$ and $x \neq -2$
    (b) $x/(x+1)$ (for $x \neq 0, -1$)
    (c) $1/(1+cx)$ (for $x \neq -1/c$ if $c \neq 0$)
    (d) $1/(1+x+y)$ (for $x+y \neq -1$)
    (e) $(x+y+2)/(x+1)(y+1)$ (for $x, y \neq -1$)
    (f) For all c, because $f(c \cdot 0) = f(0)$
    (g) For only $c = 1$, because $f(x) = f(cx)$ implies that $x = cx$, and this must be true for at least one $x \neq 0$

2. (a) $y \geq 0$ and rational, or $y \geq 1$
    (b) Rational y between -1 and 1, and all y with $|y| > 1$
    (c) 0
    (d) All w with $0 \leq w \leq 1$
    (e) $-1, 0, 1$

3. (a) $\{x : -1 \leq x \leq 1\}$
    (b) $\{x : -1 \leq x \leq 1\}$
    (c) $\{x : x \neq 1 \text{ and } x \neq 2\}$
    (d) $\{-1, 1\}$
    (e) $\varnothing$

4. (a) 2^{2y}
    (b) $\sin^2 y$
    (c) $2^{2 \sin t} + \sin(2^t)$
    (d) $\sin t^3$

5. (a) $P \circ s$
    (b) $s \circ P$
    (c) $s \circ S$
    (d) $S \circ s$
    (e) $P \circ P$
    (f) $s \circ (P + P \circ S)$
    (g) $s \circ s \circ s \circ P \circ P \circ P \circ s$
    (h) $P \circ S \circ s + s \circ S + P \circ s \circ (S + s)$

6. If
$$x = f(f(x)) = \frac{a\left(\dfrac{ax+b}{cx+d}\right)+b}{c\left(\dfrac{ax+b}{cx+d}\right)+d} \quad \text{for all } x$$

then
$$(ac + cd)x^2 + (d^2 - a^2)x - ab - bd = 0 \quad \text{for all } x$$

so
$$ac + cd = 0$$
$$d^2 - a^2 = 0$$
$$ab + bd = 0$$

It follows that $a = d$ or $a = -d$. One possibility is $a = d = 0$, in which case $f(x) = b/(cx)$, which satisfies $f(f(x)) = x$ for all $x \neq 0$. If $a = d \neq 0$, then $b = c = 0$, so $f(x) = x$. The third possibility is $a + d = 0$, so that $f(x) = (ax + b)/(cx - a)$, which satisfies $f(f(x)) = x$ for all $x \neq a/c$ (strictly speaking, we should add the proviso that $f(x) \neq a/c$ for $x \neq a/c$, which means that
$$\frac{ax+b}{cx-a} \neq \frac{a}{c}$$
or $a^2 + bc \neq 0$).

7. (a) $C_{A \cap B} = C_A \cdot C_B$
   $C_{A \cup B} = C_A + C_B - C_A \cdot C_B$
   $C_{\mathbb{R} - A} = 1 - C_A$

   (b) Let $A = \{x : f(x) = 1\}$

   (c) $f = f^2$ if and only if $f(x) = 0$ or 1 for all x; so part (b) can be applied

8. (a) Those functions f satisfying $f(x) \geq 0$ for all x
   (b) Those functions f with $f(x) \neq 0$ for all x

9. (a) y
   (b) $H(y)$
   (c) $H(y)$

10. (a)

	even	odd
even	even	neither
odd	neither	odd

(b)

	even	odd
even	even	odd
odd	odd	even

(c)

	f even	f odd
g even	even	even
g odd	even	odd

11. (a) Let $g(x) = h(x) = 1$ and let f be a function for which $f(2) \neq f(1) + f(1)$. Then $f \circ (g + h) \neq f \circ g + f \circ h$.

(b) $[(g + h) \circ f](x) = (g + h)(f(x)) = g(f(x)) + h(f(x)) = (g \circ f)(x) + (h \circ f)(x) = [(g \circ f) + (h \circ f)](x)$.

(c) $\dfrac{1}{f \circ g}(x) = \dfrac{1}{f(g(x))} = \dfrac{1}{f}(g(x)) = \left(\dfrac{1}{f} \circ g\right)(x)$

(d) Let $g(x) = 2$ and let f be a function for which $f(\frac{1}{2}) \neq 1/f(2)$. Then $1/(f \circ g) \neq f \circ (1/g)$.

12. If $f(x) = f(y)$, then $g(x) = h(f(x)) = h(f(y)) = g(y)$.

13. (a) Suppose that $x \neq y$. Then $g(x) = g(y)$ would imply that $x = f(g(x)) = f(g(y)) = y$, a contradiction.

(b) $b = f(g(b))$, so let $a = g(b)$.

14. The condition $f \circ g = g \circ f$ means that $g(x) + 1 = g(x + 1)$ for all x. There are many such g. In fact, g can be defined arbitrarily for $0 \leq x < 1$, and its values for other x determined from this equation.

15. If $f(x) = c$ for all x, then $f \circ g = g \circ f$ if and only if $c = f(g(x)) = g(f(x)) = g(x)$; that is, $c = g(c)$.

16. If $f \circ g = g \circ f$ for all g, then in particular this is true for all constant functions $g(x) = c$. It follows from the preceding problem that $f(c) = c$ for all c.

17. (a) This inequality isn't necessarily true. In fact, if $h(x) = -x$, then $f < g$ actually implies that $h \circ f > h \circ g$.

    (b) This inequality is true because $f(h(x)) < g(h(x))$ for all x.

Chapter 4

1. (a) $(2, 4)$

    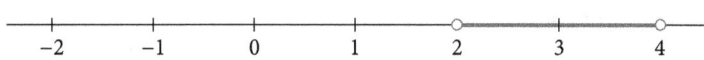

    (b) $[2, 4]$

    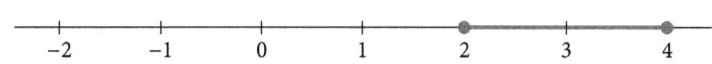

    (c) $(a - \varepsilon, a + \varepsilon)$

    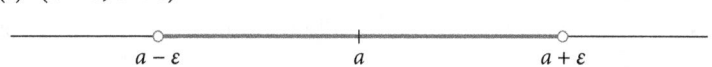

    (d) $(-\sqrt{3/2}, -\sqrt{1/2}) \cup (\sqrt{1/2}, \sqrt{3/2})$

    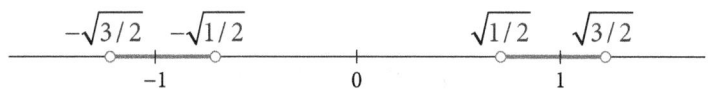

    (e) $(-2, 2)$

    (f) $\varnothing$ if $a \leq 0$;
    $\mathbb{R}$ if $a \geq 1$;
    $(-\infty, -\sqrt{(1/a)-1}] \cup [\sqrt{(1/a)-1}, \infty)$ if $0 < a < 1$

    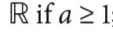 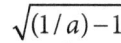

    (g) $(-\infty, 1] \cup [1, \infty)$

    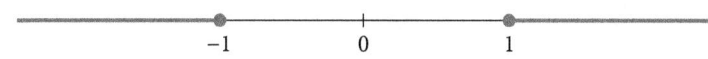

    (h) $(-1, 1) \cup (2, \infty)$

    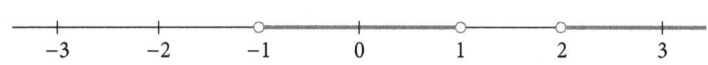

2. (a) Because $0 \le x \le b$, we have $0 \le x/b \le 1$, and $x = (x/b) \cdot b$; so choose $t = x/b$. Clearly t is the ratio at which x divides the interval $[0, b]$. The midpoint of $[0, b]$ is $b/2$.

(b) If x is in $[a, b]$, so that

$$a \le x \le b$$

then

$$0 \le x - a \le b - a$$

so that $x - a$ is in $[0, b - a]$. It follows from part (a) that for some t with $0 \le t \le 1$ we have

$$x - a = t(b - a)$$

or

$$x = a + t(b - a) = (1 - t)a + tb.$$

The midpoint of $[a, b]$ is

$$a + \frac{b-a}{2} = \frac{a+b}{2}$$

The point one-third of the way from a to b is

$$a + \frac{b-a}{3} = \frac{2}{3}a + \frac{1}{3}b$$

3. (a)

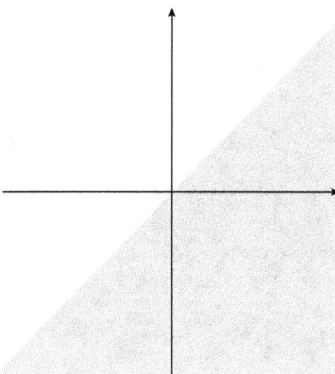

(b)

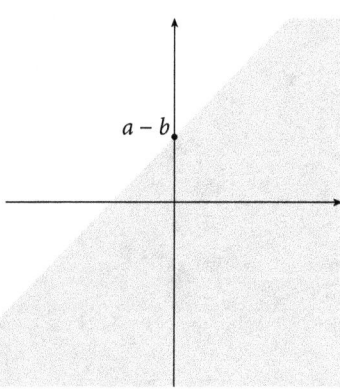

(c)

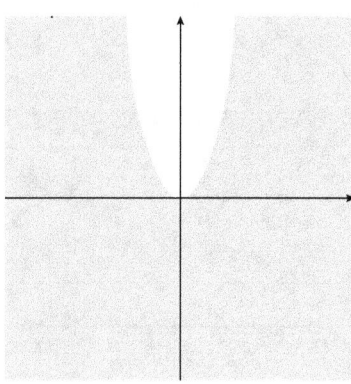

(d)

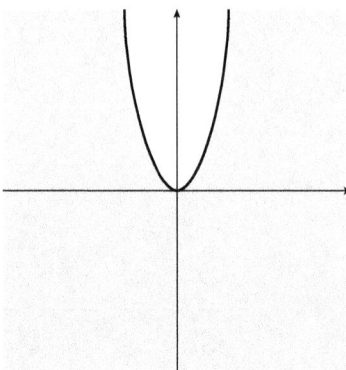

(e)

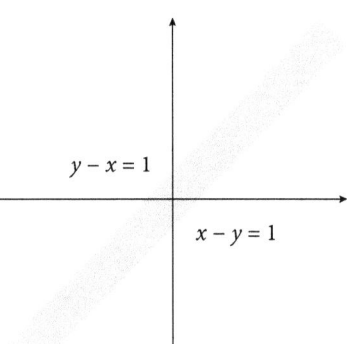

(f)

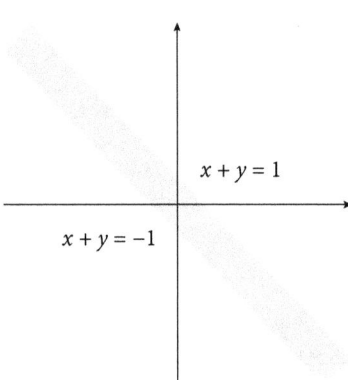

(g)

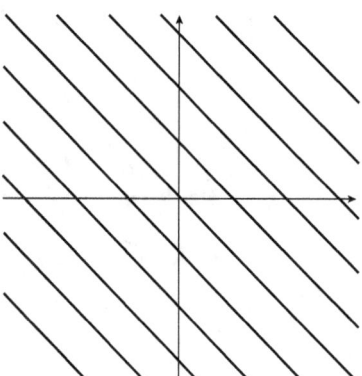

(h)

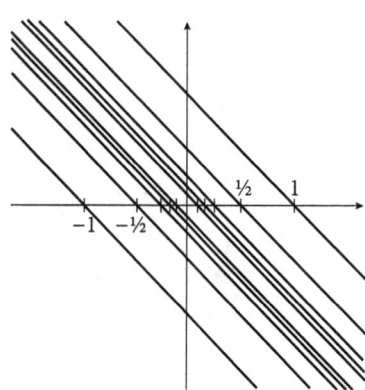

(i)

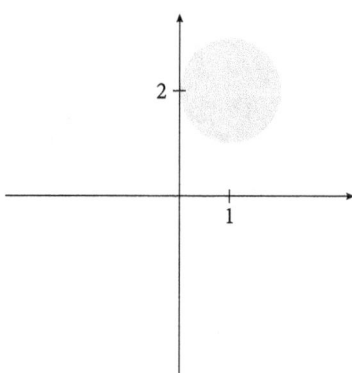

(j)

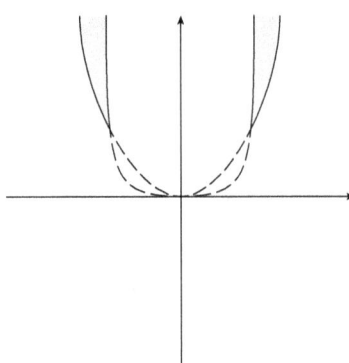

4. (a)

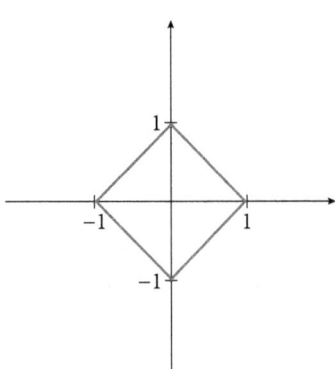

(b)

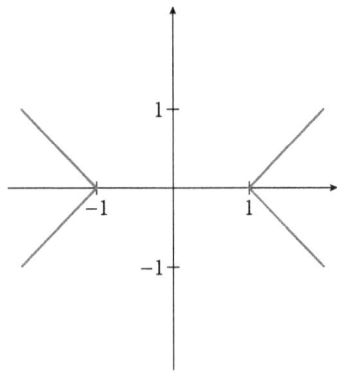

(c)

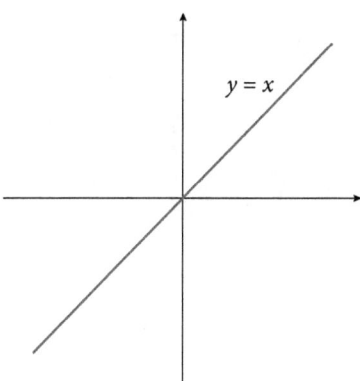

(d)

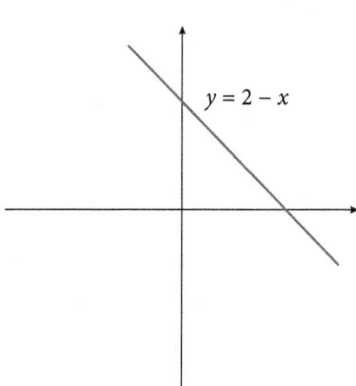

(e)

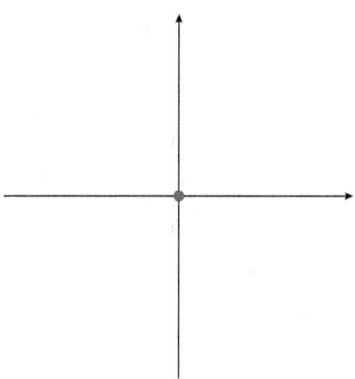

(f)

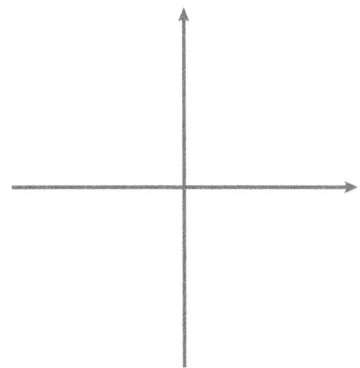

(g) $x^2 - 2x + y^2 = (x-1)^2 + y^2 - 1$

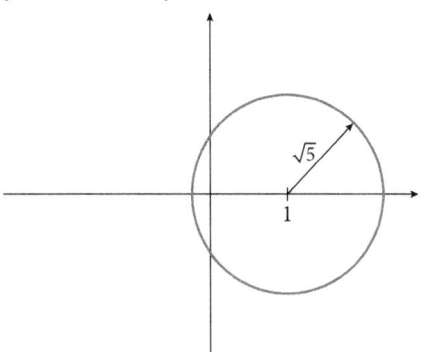

(h)

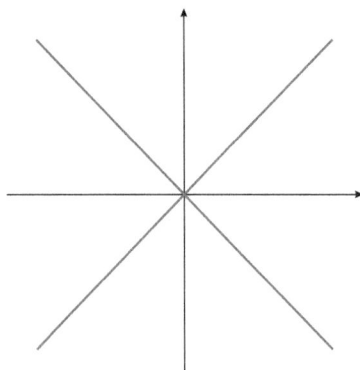

5. (a)

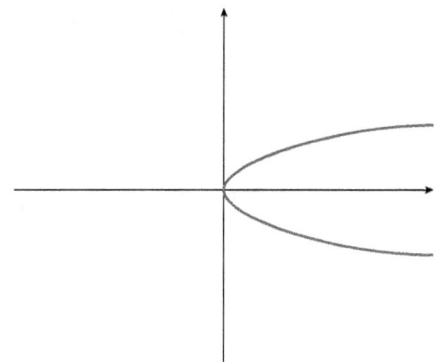

(b)

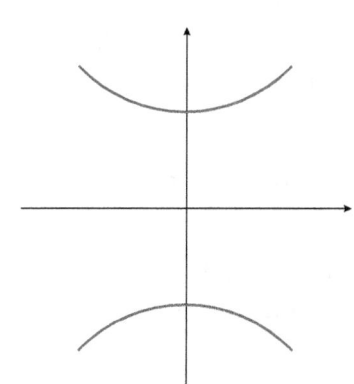

(c)

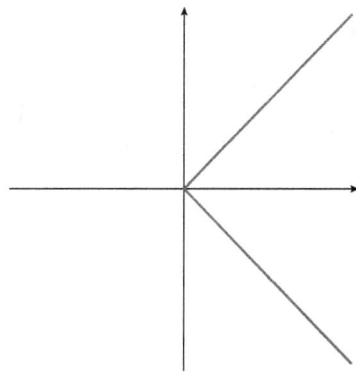

(d)

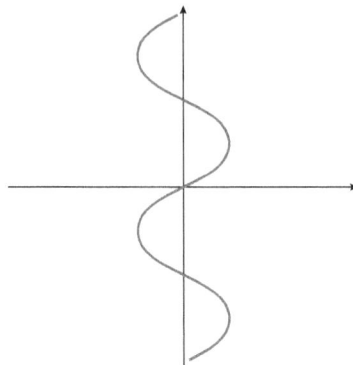

6. (a) Simply observe that the graph of $f(x) = m(x - a) + b = mx + (b - ma)$ is a straight line with slope m, which goes through the point (a, b). (The important point of this problem is simply to remember the point-slope form.)

(b) The straight line through (a, b) and (c, d) has slope $(d - b)/(c - a)$, so the equation follows from part (a).

(c) When $m = m'$ and $b \neq b'$. In that case, there's clearly no number x with $f(x) = g(x)$, while such a number x always exists if $m \neq m'$, namely, $x = (b' - b)/(m - m')$.

7. (a) If $B = 0$ and $A \neq 0$, then the set is the vertical straight line formed by all points (x, y) with $x = -C/A$. If $B \neq 0$, then the set is the graph of $f(x) = (-A/B)x + (-C/A)$.

(b) The points (x, y) on the vertical line with $x = a$ are exactly the ones that satisfy $1 \cdot x + 0 \cdot y + (-a) = 0$. The points (x, y) on the graph of $f(x) = mx + b$ are exactly the ones that satisfy $(-m)x + 1 \cdot y + (-b) = 0$.

8. (a) The angle POQ is a right angle if and only if $(PQ)^2 = (PO)^2 + (OQ)^2$.

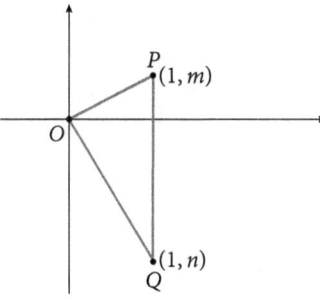

This means that
$$(m - n)^2 = m^2 + 1 + n^2 + 1$$
which is equivalent to $-2mn = 2$, or $mn = -1$. This proves the result when $b = c = 0$. The general case follows from this special case, because perpendicularity depends only on the slope.

(b) If $B \neq 0$ and $B' \neq 0$, then these straight lines are the graphs of
$$f(x) = (-A/B)x - C/A$$
$$g(x) = (-A'/B')x - C/A$$
so, by part (a), the lines are perpendicular if and only if
$$\left(-\frac{A}{B}\right) \cdot \left(-\frac{A'}{B'}\right) = -1$$
which is equivalent to $AA' + BB' = 0$. If $B = 0$ (and consequently $A \neq 0$), then the first line is vertical, so the second is perpendicular to it if and only if $A' = 0$, which happens precisely when $AA' + BB' = 0$. Similarly if $B' = 0$.

9. (a) This function is odd.

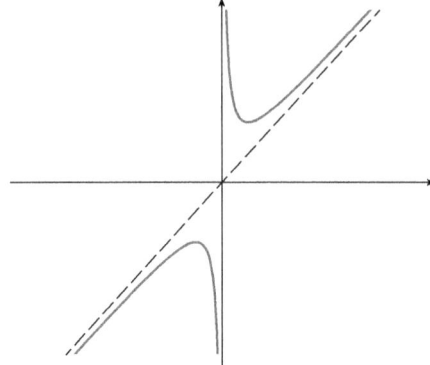

(b) This function is odd.

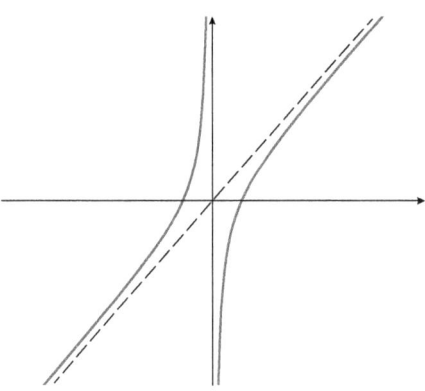

(c) This function is even.

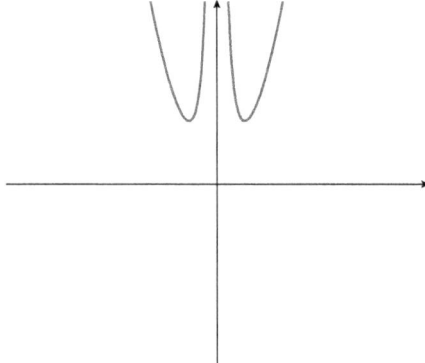

(d) This function is even.

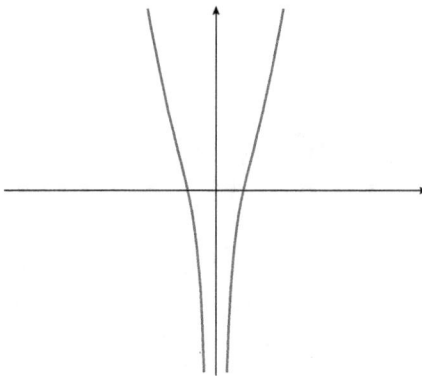

10. (a) The graph of f is symmetric with respect to the vertical axis.

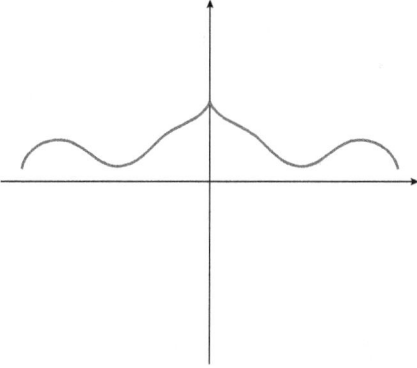

(b) The graph of f is symmetric with respect to the origin. Equivalently, the part of the graph to the left of the vertical axis is obtained by reflecting first through the vertical axis, and then through the horizontal axis.

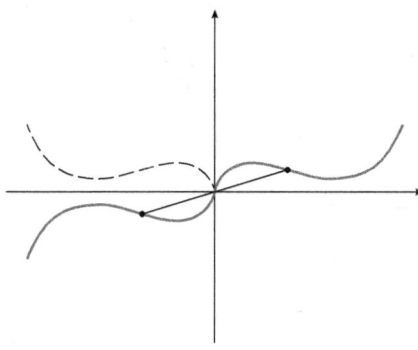

(c) The graph of f lies above or on the horizontal axis.

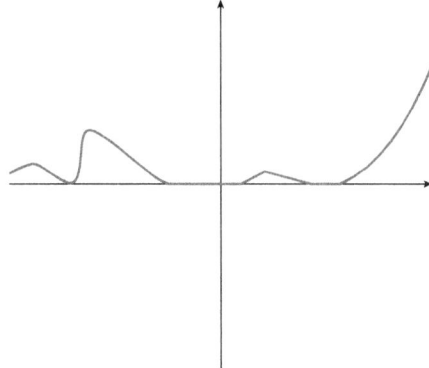

(d) The graph of f repeats the part between 0 and a over and over.

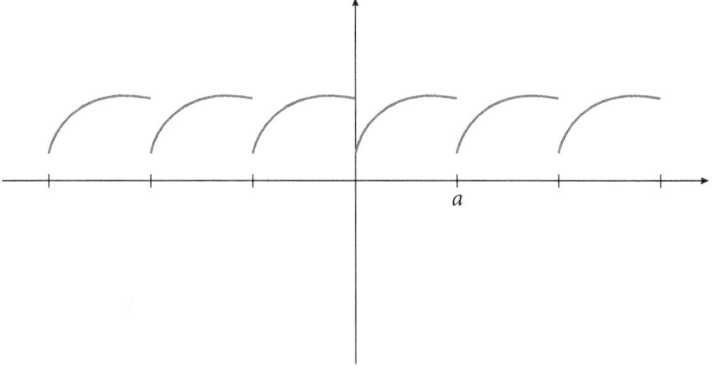

11. When n is odd, the domain of f is $\mathbb{R}$, but when n is even, the domain of f is $[0, \infty)$.

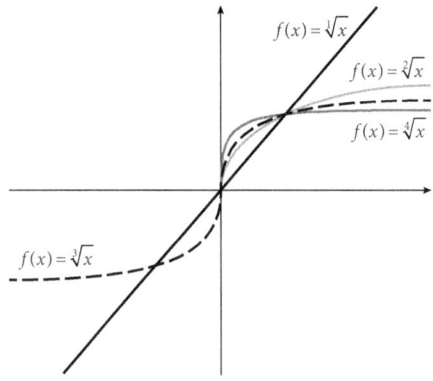

12. The graphs of $f(x) = |x|$ and $f(x) = |\sin x|$ contain "corners".

(a)

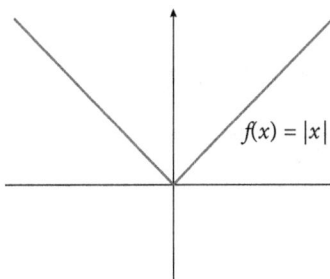

(b)

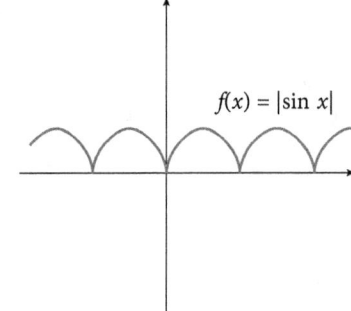

13. (a) The graph of g is the graph of f moved up c units.

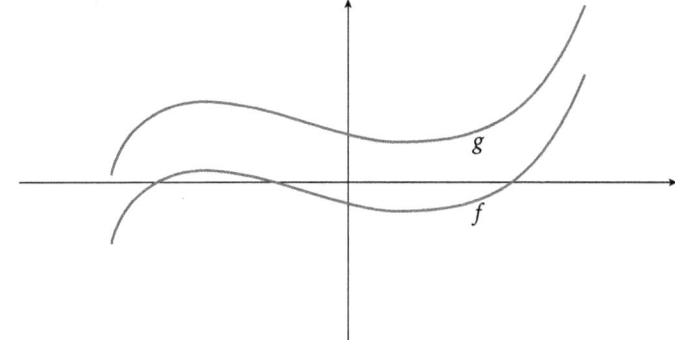

(b) The graph of g is the graph of f moved c units to the *left* (if c > 0).

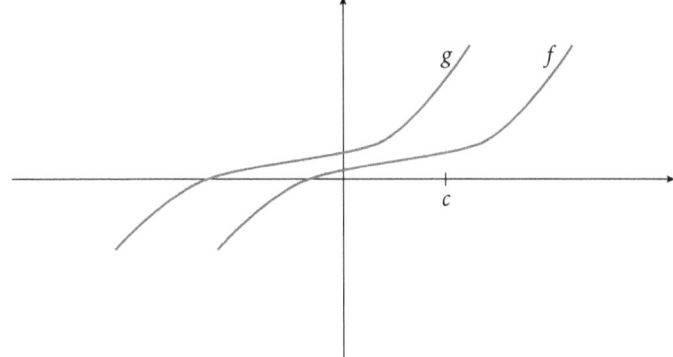

(c) The height of the graph of f is multiplied by a factor of c everywhere. If c = 0, then g = 0. If c > 0, then distances from the horizontal increase in the same direction. If c < 0, then distances are increased but directions are changed.

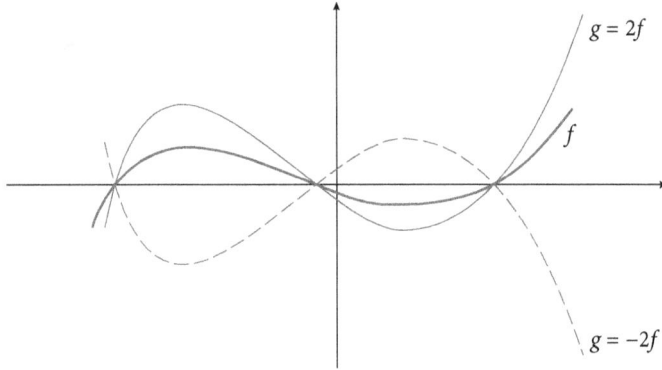

(d) The graph of f is compressed by a factor of c if $c > 0$. If $c < 0$, then compression is combined with reflection through the vertical axis. If $c = 0$, then g is a constant function, $g(x) = f(0)$.

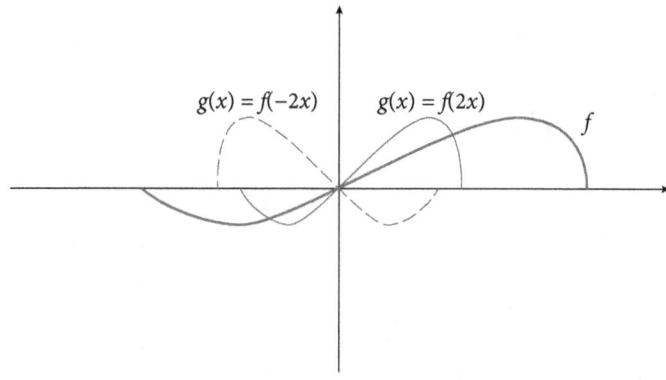

(e) Everything that happens far out happens near 0, and vice versa, as illustrated by the graph of $g(x) = \sin(1/x)$.

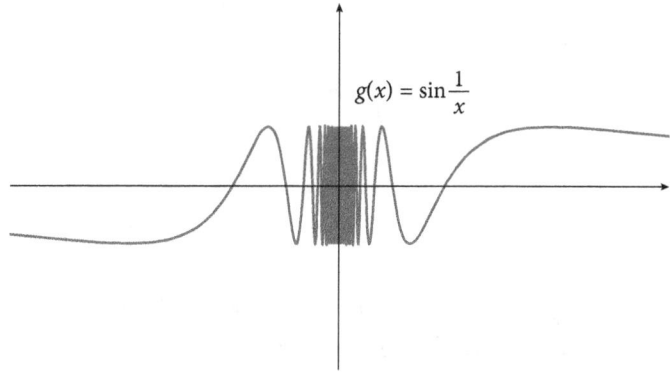

(f) The graph of g consists of the part of the graph to the right of the vertical axis, together with its reflection through the vertical axis.

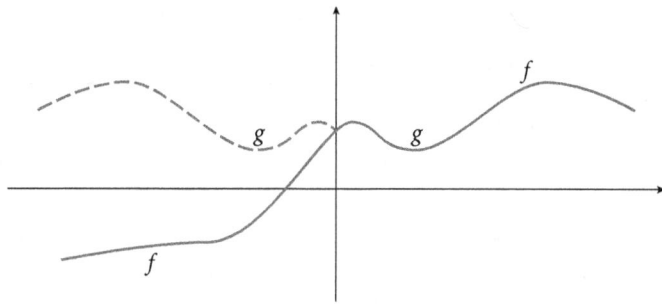

(g) The graph of g is obtained by flipping up any parts of the graph of f which lie below the horizontal axis.

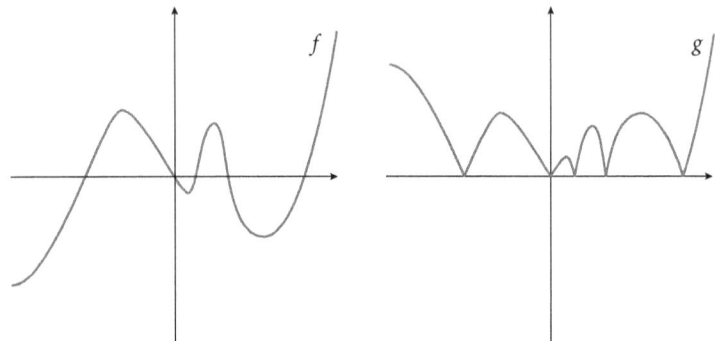

(h) The graph of g is obtained by "cutting off" the part of the graph of f which lies below the horizontal axis.

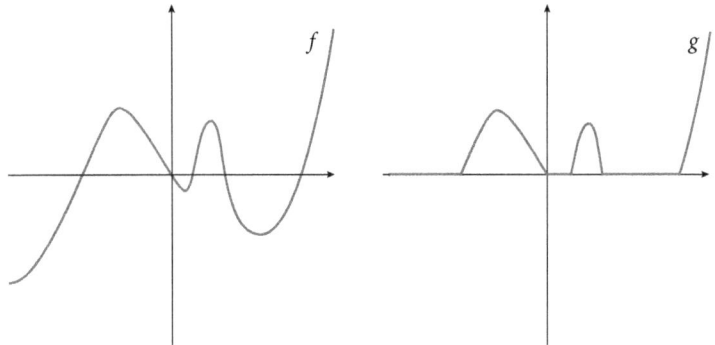

(i) The graph of g is obtained by "cutting off" the part of the graph of f which lies above the horizontal axis.

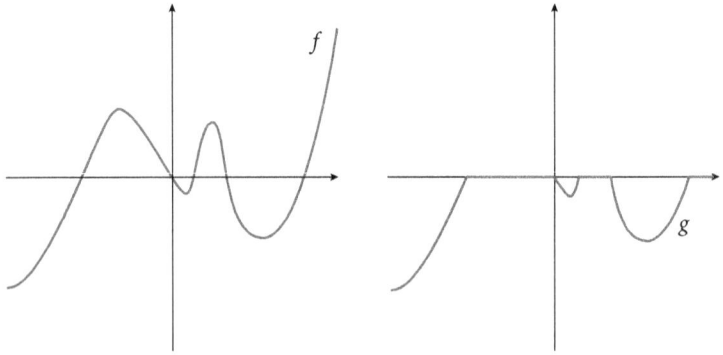

(j) The graph of g is obtained by "cutting off" the part of the graph of f which lies below the horizontal line at height 1.

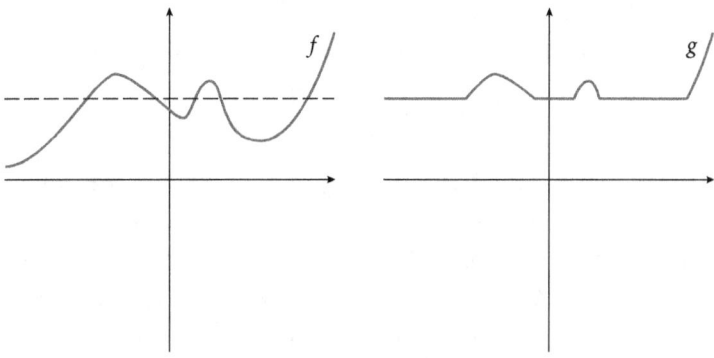

14. Because
$$f(x) = ax^2 + bx + c$$
$$= a\left(x^2 + \frac{b}{a}x + \frac{c}{a}\right)$$
$$= a\left[\left(x + \frac{b}{2a}\right)^2 + \left(\frac{c}{a} - \frac{b^2}{4a}\right)\right]$$

the graph looks like the following figure.

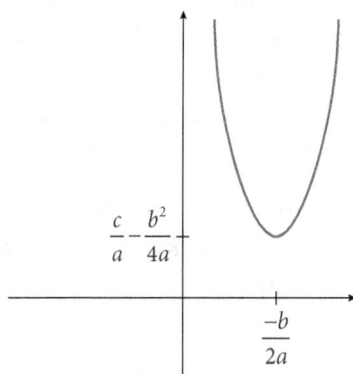

15. Suppose $C = 0$, so that we have the equation

$$Ax^2 + Bx + Dy + E = 0$$

If $D \neq 0$, then this is equivalent to

$$y = -\frac{A}{D}x^2 - \frac{B}{D}x - \frac{E}{D}$$

so the set of all (x, y) satisfying this equation is the same as the graph of $f(x) = (-A/D)x^2 - (B/D)x - (E/D)$, which is a parabola, by the preceding problem. If $D = 0$, then we have the equation $Ax^2 + Bx + E = 0$, $(A \neq 0)$, which may have zero, one, or two solutions for x; in this case the set of all (x, y) satisfying the equation is either $\emptyset$, one straight line, or two parallel straight lines. Similarly, if $A = 0$, then we again have a parabola (compare Problem 5(a)). When A, $C \neq 0$ we can write the equation as

$$A\left(x + \frac{B}{2A}\right)^2 + C\left(y + \frac{D}{2C}\right)^2 = F$$

for some F.

When $A = C > 0$ we have a circle, unless $F = 0$, in which case we have a point (a "circle of radius 0"), or $F < 0$, in which case we have $\emptyset$. In general, when $A, C > 0$ we have an ellipse not necessarily centered at the origin (or a point, or $\emptyset$). There's no need to consider separately the case $A, C < 0$, because we have the same situation, replacing F by $-F$.

When A and C have opposite signs we have a hyperbola for $F \neq 0$ (the direction it points depends on the signs of A, C, and F). For $F = 0$ we have the equation

$$x + \frac{B}{2A} = \pm\sqrt{\frac{-C}{A}}\left(y + \frac{D}{2C}\right)$$

which gives two intersecting lines (a "degenerate hyperbola").

16. (a)

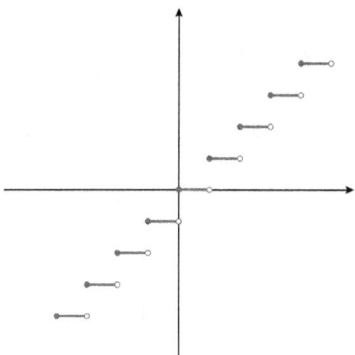

(b)

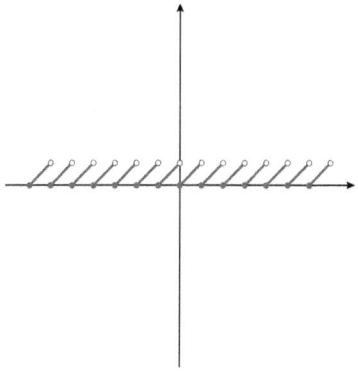

(c)

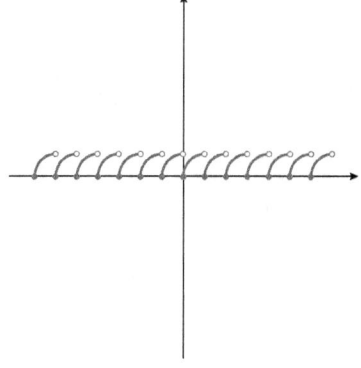

(d)

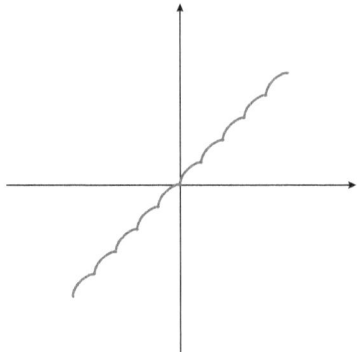

17.

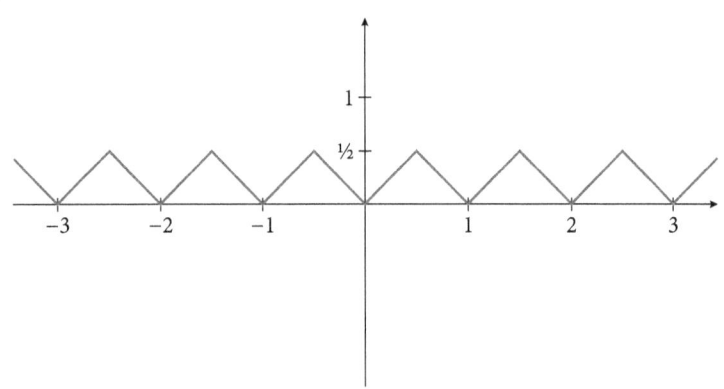

18. (a) Note that different scales have been used on the two axes.

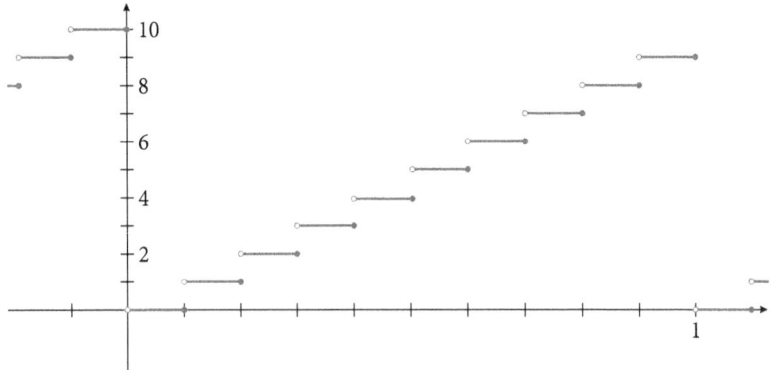

(b) The graph of f is similar to the graph in part (a), except that there are ten sets of ten steps between n and $n + 1$.

19. (a) The square of the distance from (x, x^2) to $(0, \frac{1}{4})$ is

$$(x-0)^2 + \left(x^2 - \frac{1}{4}\right)^2 = x^2 + x^4 - \frac{x^2}{2} + \frac{1}{16}$$

$$= x^4 + \frac{x^2}{2} + \frac{1}{16}$$

$$= \left(x^2 + \frac{1}{4}\right)^2$$

which is the square of the distance from (x, x^2) to the graph of g.

(b) The point (x, y) satisfies this condition if and only if

$$(x - \alpha)^2 + (y - \beta)^2 = (y - \gamma)^2$$

or

$$x^2 - 2\alpha x + \alpha^2 + y^2 - 2\beta y + \beta^2 = y^2 - 2\gamma y + \gamma^2$$

or

$$y = \left(\frac{1}{2\beta - 2\gamma}\right)x^2 + \left(\frac{\alpha}{\gamma - \beta}\right)x + \left(\frac{\alpha^2 + \beta^2 - \gamma^2}{2\beta - 2\gamma}\right)$$

This solution works only for $\beta \neq \gamma$, which is simply the condition that P is not on L. If P is on L, then the solution is the vertical line through P.

Index

Symbols

× (cartesian product) 5
∘ (composition) 73
− (difference) 14
∈ (element of) 1
∅ (empty set) 2, 51, 84
! (factorial) 52
∞ (infinity) 85
∩ (intersection) 14
ℕ (natural numbers) 2, 49
∉ (not an element of) 1
⊄ (not a subset of) 8
ℚ (rational numbers) 4, 55
ℝ (real numbers) 2, 55, 84
{} (set members) 1
⊆ (subset of) 8
∪ (union) 14
ℤ (integers) 2, 55
ε (epsilon) 84
Σ (summation) 53

A

absolute value 47, 60
addition 35
associative law for addition (P1) 36, 43
associative law for multiplication (P5) 38, 43
axes (horizontal and vertical) 86

B

binomial coefficient 62

C

cardinality (sets) 2
cartesian power (sets) 8
cartesian product (sets) 5
circles 99
closure under addition (P11) 44
closure under multiplication (P12) 44
commutative law for addition (P4) 38, 43
commutative law for multiplication (P8) 39, 43
complement (sets) 16
composition (functions) 73
constant functions 71, 87
coordinates 86
coordinate system 86

D

degree (polynomial functions) 70
difference (sets) 14
distance between points 89
distributive law (P9) 40, 43
division 39
 by 0 39, 46, 58
domain (functions) 67, 76

E

elements (sets) 1
ellipses 99
empty set 2, 51, 84
even functions 81, 107
even numbers 55, 63

165

existence of additive inverses (P3) 37, 43
existence of a multiplicative identity (P6) 38, 43
existence of an additive identity (P2) 37, 43
existence of multiplicative inverses (P7) 38, 43

F

factorial 52
factorization 42
functions
  circle 99
  composition of 73
  concept of 65
  constant 71, 87
  defined 65, 74
  domain of 67, 76
  ellipse 99
  even 81, 107
  examples of 65
  hyperbola 101
  largest-integer (floor) 108
  linear 89, 105
    point-slope form 90, 105
  naming 67
  nearest-integer 108
  notation 67
  odd 81, 107
  ordered pairs 75, 77
  oscillating 95
  parabola 91, 108, 109
  periodic 107
  polynomial 70, 93
  power 92, 107
  product of 71
  quotient of 71
  rational 70, 94
  reasonable vs. unreasonable 103
  sine 96
  sum of 71
  value of 67

G

geometric figures 99, 108
  circles 99
  ellipses 99
  hyperbolas 101
graphs
  accuracy of 92, 97
  and geometric figures 102
  defined 86
  proofs involving 88
  rotating 102
  vertical-line test 90

H

horizontal axis 86
hyperbolas 101

I

indexed sets 20
inequalities 44
infinity 85
integers 2, 55
intersection (sets) 14
intervals 5, 84
irrational numbers 5, 55, 63, 83, 98

L

largest-integer (floor) function 108
linear functions 89, 105
  point-slope form 90, 105

M

mathematical induction 49, 62, 63
multiplication 38, 42
  of negative numbers 41

N

natural numbers 2, 49
nearest-integer function 108
negative numbers 41, 44
null set. *See* empty set
number line. *See* real line
numbers
  absolute value 47, 60
  even 55, 63
  factorial 52
  integers 2, 55
  irrational 5, 55, 63, 83, 98
  natural 2, 49

negative 41, 44
odd 55, 63
positive 44
rational 4, 55, 63, 83, 98
real 2, 55, 84
square root 48, 55

O
odd functions 81, 107
odd numbers 55, 63
ordered pairs 5, 75, 77
origin
   plane 86
   real line 83
oscillating functions 95

P
parabola functions 91, 108, 109
periodic functions 107
perpendicular lines 106
points
   distance between 89
   representing numbers as 84
point-slope form 90, 105
polynomial functions 70, 93
positive numbers 44
power functions 92, 107
power sets 12
principle of complete induction 52
properties
   associative law for addition (P1) 36, 43
   associative law for multiplication (P5) 38, 43
   closure under addition (P11) 44
   closure under multiplication (P12) 44
   commutative law for addition (P4) 38, 43
   commutative law for multiplication (P8) 39, 43
   distributive law (P9) 40, 43
   existence of additive inverses (P3) 37, 43
   existence of a multiplicative identity (P6) 38, 43
   existence of an additive identity (P2) 37, 43
   existence of multiplicative inverses (P7) 38, 43
   trichotomy law (P10) 44
Pythagorean theorem 89

R
rational functions 70, 94
rational numbers 4, 55, 63, 83, 98
real line 83
   intervals 5, 84
real numbers
   and infinity 85
   as coordinates 86
   as points 84
   decimal expansion 83, 108
   irrational 5, 55, 63, 83, 98
   rational 4, 55, 63, 83, 98
   representing geometrically 83
   set of 2, 55, 84
recursive definition
   of exponentiation 63
   of factorial 52
   of summation 54
Russell's paradox (sets) 27

S
sets
   cardinality of 2
   cartesian power of 8
   cartesian product of 5
   complement of 16
   defined 1
   difference of 14
   elements of 1
   empty 2, 51, 84
   equal 1
   finite 1
   indexed 20
   infinite 1
   integers 2, 55
   intersection of 14
   intervals 5, 84
   natural numbers 2, 49
   notation 1, 3
   ordered pairs 5
   power sets of 12

rational numbers 4, 55
real numbers 2, 55, 84
Russell's paradox 27
subsets of 8
union of 14
universal 16
Venn diagrams 17
well-ordering principle 25, 51
Zermelo–Fraenkel axioms 28
sigma notation 53
sine function 96
slope (linear functions) 88
square root 48, 55
subsets 8
subtraction 37, 38
summation (Σ) 53

T
trichotomy law (P10) 44

U
union (sets) 14
unit circle 99
universal set 16

V
Venn diagrams (sets) 17
vertical axis 86
vertical-line test 90

W
well-ordering principle (sets) 25, 51

Z
Zermelo–Fraenkel axioms 28
zero point. *See* origin

www.ingramcontent.com/pod-product-compliance
Lightning Source LLC
Chambersburg PA
CBHW052317220526
45472CB00001B/154